to Cathy—
who is my old friend now—
Bob
December '86

S·E·A
C·H·A·N·G·E·S

Robert Kotlowitz

NORTH POINT PRESS : SAN FRANCISCO
1986

Copyright © 1986 by Robert Kotlowitz
Published in the United States of America
Library of Congress Catalogue Card Number: 86-60998
ISBN: 0-86547-252-1

Grateful acknowledgment is extended
to the National Endowment for the Arts, for a grant
that helped to make the writing of this book possible,
and to Alfred I. Coplan and RelStores, Inc., for permission to use the slogan, "Marriages are made in heaven . . .
engagements are made at M. Gordon & Son."

"Stouthearted Men" by Oscar Hammerstein II
and Sigmund Romberg. Copyright © 1927 (renewed)
Warner Bros. Inc. All rights reserved. "Smoke Gets in
Your Eyes" by Jerome Kern and Otto Harbach.
Copyright © 1933, T. B. Harms Company. Copyright
renewed (c/o The Welk Music Group, Santa Monica,
California 90401). International copyright secured.
All rights reserved. Used by permission.

FOR MY MOTHER, DEBBY KOTLOWITZ

Part One

Chapter One

Summer that year held into September, golden and hot, surfacing a wave of unexpected optimism that took most people by surprise; but like summer itself, like the long baking days that hung on precariously for a few weeks more, it didn't last. Nothing lasted then, in any case; everybody understood that. What next, they all asked each other in fierce voices, what will happen? By early October, as the sharp fall colors swept steadily south, working their way through Europe like a glacial tide, the latest Bavarian party rally had mesmerized the entire nation again. A familiar trance fell across the country. Everyone was bewitched, regardless of politics. Children had silent crying fits. Old people raged in their sleep against the prospect of missing out on the future. Certain farmers in the Palatinate, enthralled by the news reports and the promises from Nurnberg, claimed to see flaming stars shooting across the Rhenish sky, against all scientific predictions. A few evenings later, another band of country citizens noted that the autumn moon rose two hours early north of Gottingen, near the Harz mountains, and on the same night, in Berlin, distant comets were seen flying boldly from Prussia to the east, towards the wheatfields of Poland and the Ukraine. Within days, however, the heavens over Germany returned to normal. Peculiar phenomena vanished. Days fulfilled the requisite hours; dusk fell earlier each evening, as expected; hour by hour, it grew chillier; and in Frankfurt-am-Main, huddled grey and humpbacked on the great river, where it had rained for seventy-two hours without a break, the old habits soon tried to take over again at the Vogel dinner table and failed.

Small talk circled warily around each course, looking for an opening. Ordinary banter virtually disappeared as the family began to withdraw into itself. Julius, the father, set the style, Elsa, his wife, helped to confirm it. They were naturally shy anyway. The meal itself, never really joyous even in the best of times, became a solemn evening chore from which they all flinched, and Julius, Elsa, and their two sons mostly sat around the table now in an unhappy silence that was both secretive and irritable. Outside, the Nurnberg roar had reached the city in epic surges and was still echoing along Frankfurt's broad avenues and would go on echoing, but the Vogels, stiffly seated face-to-face in their dining room off the Mendelssohnstrasse, had clearly heard enough.

Only Leopold stayed at the sharp old pitch, his normal edgy one, trying to raise the others to his own nervous, often precocious level by cracking politically sensitive jokes. "The Germans represent a medical miracle," he shouted one night over the soup, while Julius and Elsa stared at their plates. "They're able to walk around upright in spite of having a broken backbone." Not such a great joke, maybe (nobody laughed), not even adequate, but a joke, nevertheless, and jokes helped to cheer Leopold up. He loved telling them, swallowing his words from excitement, watching his family for the effect they might have, waiting to be acknowleged; but at the sound of his eager voice, a little too eager, perhaps, as always, Julius and Elsa Vogel turned their heads away and pretended to be thinking of other things. Neither wanted to pay attention; neither had the patience any longer. The Vogels were feeling prickly; their household was on steady alert; Leopold's jokes might turn out to be dangerous. There was also an unspoken feeling that laughter was somehow unfitting these days, that they might be punished for it. Each night, Leopold's family couldn't wait for dinner to end because they no longer enjoyed talking to each other. They had to get away from the table. Terrible things were happening to them and to others—"The passing excesses of the day," Kurt Vogel described them, with a supercilious (and typical) irony that perfectly masked his real feelings. But no one at the Vogel dinner table shared true feelings anymore. True feelings had begun to vanish with everything else. They told each other lies even when the truth was harmless; common, trivial, pointless lies; ongoing and unchallenged. The habit of truth was broken. Like many of their friends, who also chose to remain silent among themselves, the Vogels had already gone into hiding, they wore a kind of spiritual camouflage, in public and at home, without acknowledging it to themselves.

There were really no terms to explain the present, Julius Vogel understood, looking into his sons' eyes—and his wife's, as well—at dinner, night after night. He would not even try to explain it to himself. The matter of his insurance business was only a part of it. That fall, Julius had started to dream again that the business might eventually make them all rich one day. It was an old dream, an innocent conventional one designed to assure the future; a father's ongoing foolish dream; but almost overnight, without warning, Julius lost most of his Aryan clients, his main source of profit, by official government decree. This was actually a subclause of the new Nurnberg Laws. Within a week, more than sixty percent of the premiums owed the Vogel agency was forcibly transferred to another, legally acceptable firm, which simply gathered them in as though Julius and his business had never existed. It was that simple, that direct. There was no chance to protest, no so-called redress; the State had proclaimed it. Julius even had to help in the transfer himself, working blindly until midnight with one of his own clerks, sharing most of the paperwork. Together, as they worked, they ate overstuffed liverwurst sandwiches prepared by Elsa Vogel and drank hot coffee brought in from a nearby cafe where they often lunched, and premium by premium, policy by policy, client by client, efficiently got rid of more than half the business in two long nighttime sessions. The procedure almost bankrupted Julius on every level, as intended, and before long, as a quiet panic took hold of him, something new, never experienced before, something peculiarly isolating, the Vogel family lost a large share of its regular earned income, bringing on an immediate financial crisis in the household, as well as other crises peculiar to each one of them.

As a result of the New Laws, for example, Elsa Vogel stopped menstruating almost immediately following the unabridged publication, without mercy, of each Nurnberg decree in the *Frankfurter Zeitung*. Julius was not surprised. His wife had a long history of neurasthenic ailments that came and went, apparently as she willed them—bruised vocal nodes, unexplained body rashes, sudden nosebleeds—and this one followed the traditional patterns until it became clear that it was more than one of her passing fancies. There was no effective treatment for this illness, they learned from the family physician, no useful medication or surgical technique that might act as a cure. Nor would time, in all probability, have a transforming effect on such a drastic symptom. It seemed that Elsa Vogel had made her comment on the question of Aryans and non-Aryans, and there was no contradicting it. Dr. Schwabacher offered his opinions in matter-of-fact terms; they were all old friends, going back to gymnasium. Yes, it was premature, he said, Elsa, of course, was not yet ready for a normal menopause. But who could tell about these things? How could anyone be sure? And yes, he went on when pressed, life was strange, they all knew that, but it was sometimes strange in beautiful and curiously fitting ways, was it not? Such things came with the times, he finished, without much heart. Then, turning aside at the accusing look on Julius's face, Dr. Schwabacher unhappily added Elsa Vogel to his growing collection of twentieth-century case histories.

In the new round of things, it was Leopold's turn next. One Sunday afternoon, soon after his father lost the business, a day after Dr. Schwabacher's latest visit to Elsa Vogel, Leopold was out on his own, looking for company. As usual, he had wandered as far as the Schillerplatz, in the center of town. (If Leopold didn't go to the movies on Sunday, he went to the Schillerplatz, where he could buy pastries at the Cafe Bauer with his allowance money and, moving through the crowd, pretend that he was part of the great world of Frankfurt.) Two things happened this time. Early on, hanging around the poet's memorial with his hands in his pockets and his foot resting jauntily on the base of the statue, he was snubbed by a pair of old school friends who were not Jewish. "Hello," he called, as they walked past him without answering, staring rigidly ahead as though they had made a pact with the future. Karl Ebert, toothy son of a local pharmacist, arm in arm with Johann Bader, whose mother came from Trieste; both silent, mouths shut. It was a new experience for Leopold Vogel.

Then, later in the afternoon, as he was starting for home (still without company), he ran into a pack of adolescent strangers from another part of town who, without warning, began to knock him around. He hardly had time to think as they first surrounded him, then began to force him into an empty side street, crying "Hep! Hep!" in high-pitched voices. In the Schillerplatz, behind them, the thin crowd scurried by. It was almost evening. No one interfered. No one objected. It was a children's brawl. First, the strangers made a little circle around him, pushing him into each other's arms. It was like a game they all played at recess in school. As they caught him, they kept laughing as though to keep up their courage. Then one of them pulled Leopold's ears. He fell when they began to trip him, then fell again to save himself. A full minute passed as he lay there on the cobblestones in a terrible hovering silence. While he waited, Leopold caught a glimpse of the bald autumn sun hanging at the end of the street, just above the chilled line of the horizon. There was hardly a sound, except for the traffic in the Schillerplatz. The rest of the world had disappeared. With any luck, they would leave him there to die. One of the boys standing over him had a round flat face, like one of Elsa Vogel's dinner plates. His face had scabs everywhere. "Up!" he ordered, pulling Leopold by two fingers. Leopold tried to stay on his feet. "Hep!" a voice shouted from behind him. The strangers began to shove him around the circle again. He was able to count five of them, besides the one with the scabbed face. In the end, they finished him off by urinating on his expensive new knickers. They did it as though it was their duty, a Sunday obligation for good Lutherans out for a Sabbath stroll. Then they ran off, shouting "Hep!" in march-time. Minutes passed while dismay and shock froze Leopold Vogel. His hand was bleeding, sweat ran from his armpits. In his terror, he had not struck a single blow in defense of himself. He hadn't even made a threatening fist. He stood there shamefaced in his soaking corduroys, in a rage at himself and the stink he made, his heart beating wildly on the streets of Frankfurt. This too was a new experience.

Later that night, after safely sneaking into the house to change his clothes in secret, he was shrouded into sleep by a sense of cowardice that was also new to him. By morning, after a sleep without dreams, it had become a terror in itself. He could see it staring at him eye to eye from the bathroom mirror when he brushed his teeth before breakfast, facing himself in the clear neutral light. Where had that thin-lipped, sodden face come from, he asked himself, looking down at the floor with sudden shame. It didn't belong to Leopold Vogel, it was not a face he had ever known. Then, thrusting himself aside distractedly, he hurried into the dining room for a roll and hot coffee before leaving for school. In class that day, Leopold made himself into something of a hero by telling the story, with certain creative distortions, to anyone who would listen. He discovered that he couldn't contain the narrative, it seemed to have a life of its own, like one or two of his jokes. By the end of the day, he was even making some of his listeners laugh. Resiliency and nerve carried him through. Shame began to dissipate. It was how he always took heart, it was habit, but he didn't tell his family about the episode until a few days later, when he finally had to explain the condition of his knickers to his mother.

And finally, around the same time—and bound tightly to all these events—Kurt Vogel, older and tougher than his brother, Leopold, made secret application at the British consulate for a UK visa. He would get out of Frankfurt, he decided, flee his fat old *Deutsches Burg* and head for London. To help assure the peace at home, he kept the decision to himself. His parents would have to wait for the news. He would give them plenty of notice, in due time. For the moment, they had their hands full with themselves and their younger son.

Chapter Two

In America, meanwhile—in a northwest corner of Baltimore, Maryland—Florence and Max Gordon pursued their adoption of a German refugee boy, for whom they had waited a full year. Leopold Vogel, of the expensive new knickers and the questionable jokes, was to be this boy. His parents had made the decision on his behalf months before, and he had agreed, perhaps a little too quickly in their eyes. They had then pulled back, changed their minds, and, typically, changed their minds again. But he was going. That was settled now. America or the Schillerplatz: how could they hesitate?

Applications had been exchanged with the authorities, proofs of identity shared by both families, character references gathered from friends and community leaders. Leopold was examined and pronounced fit by three doctors: internist, dentist, and psychiatrist (a touch of excessive albumin was discovered in his blood by Dr. Schwabacher, the dentist thought his teeth needed straightening, nothing serious, and based on a hurried interview with the psychiatrist it seemed clear that there were no obvious personality disorders); while across the sea the Gordons, in vouching that Leopold would not become a public charge in America, had to show proof that Max Gordon made enough money to support him. This they did, just barely; but barely, in this case, was good enough. The Gordons were financially sound, within limits. They also promised—in fact, demonstrated with floor plans specially drawn by Max himself—that Leopold would have his own room in the Gordon household. Through all this, Florence and her husband, who owned a jewelry store in downtown Baltimore, were interviewed several times at the Jewish Welfare Agency offices, which handled such cases with its equivalents in Frankfurt and other cities. It was all part of the routine, one of the new exercises of the day, almost predictable by now. At first they were nervous in these exchanges, tending to contradict each other impulsively over the most ordinary matters, but soon, as the agency social worker and the official surroundings grew familiar, they began to feel comfortable enough within the process. The third time around, they already knew how to manipulate the questioning on their own behalf; and after another week or so, during which they busily varnished their undeniable quirks to the point of invisibility, they were judged acceptable as future foster-parents. It helped that they went easy on the fact that they had been trying unsuccessfully for years to conceive a second child of their own—an authentic biological Gordon—to follow their first, who happened to be a girl named Adele. In fact, they never mentioned it to the social worker, as Max's sister, Clara, had insisted in pre-interview rehearsals at home. "Don't be a fool," Clara told her sister-in-law. "I know how these things work. Act as if everything is perfect. Don't beg any questions." They did what she told them, begging no questions, concentrating instead on their only child's health and psychic well-being, and on their own as well, and, to no one's surprise, Clara's advice worked. (Clara's advice always worked.) The Jewish Welfare Agency put a double check next to their names: top priority. They would have their son; the Vogels would lose theirs. So all the official papers, which had to travel back and forth between Baltimore and Frankfurt many times, all the assurances, promises, proofs, and quasi-legal oaths, were eventually stamped with the correct government seals and appropriate national symbols, awaiting final disposition.

Nevertheless, the case of Leopold Vogel (RF-LV-217, in the Agency files), like all such cases, dragged on with almost malicious slowness. Would it ever happen, they asked themselves over and over, Gordons and Vogels alike. Would he be allowed to leave one continent and sail for the other? While they waited, both families remained nervous. Everyone was testy, high-strung. There were helpless domestic set-tos in Frankfurt and Baltimore between husband and wife, parents and children, emotional firestorms that flashed for a few moments, burned themselves out, and left everyone exhausted. Julius Vogel told himself that there was probably no other way—so did Max Gordon—given the demands of the transaction. For neither man, German or American, was there any precedent for all this. They could only wait it out, persuading themselves that they were in control, and try to keep their families in reasonable emotional order.

In the meantime, fall moved interminably on. Leaves splattered the sidewalks of Frankfurt and Baltimore, the bleached sun distanced itself in a smoky sky. It grew cold everywhere, then warm again, but in the end, of course, the chill season finally held. Other things happened as the weeks ground ahead, many things, in fact, some of them within the Gordon family in America. Around Halloween Eve that year, for instance, towards dusk of a flaming orange day, Adele Gordon let herself be held, and more than held, touched and touched again and something more by a boy who lived down the street on Fairfax Road in a rowhouse just like the Gordons'. The boy's name was Billy Brent and Adele was his wholehearted partner—not a victim, never a victim—having consciously provoked the incident by suggesting to Billy on the way home from school that he was looking kind of peaked these days. That was her word: peaked. Did Billy feel all right? He was very pale. Everybody had noticed it. Was looking peaked, Adele inquired without embarrassment, maybe a sign of being horny? That's what somebody in her class had told her, she claimed, fidgeting with her books. But it was a fib; nobody had told her anything; she had made the whole story up.

As she spoke, she offered Billy her straight-in-the-eye-I'm-not-afraid-of-anything look and a little half-smile that gave it all away, then glanced involuntarily down at Billy's crotch. It didn't take long after that for the two of them to head for the alleyway and the Kates's garage, next door to the Gordons', which was used by the Kates family to store junk. On the way, they were careful not to knock over any empty garbage cans and alert the neighborhood to what was going on.

At first, inside the garage, there was not much more than some hurried pushing around in the dark. Here, they both called, in suddenly small, hungry voices, considerately guiding each other. Adele tripped over an old inner tube. Billy caught her. They touched and touched again. Soon, for Billy and his practiced hands, Adele, shivering among the mouse droppings, let go a sigh that heaved through the tin garage and came echoing softly back at them from the corrugated walls. For a moment she hardly recognized the sound of her own voice. Jee-sus, Billy whispered. Hold on. He couldn't believe the power of it. Wait for me, he breathed. But by then, Adele could hear nothing, she was so far beyond herself. It was all over in five minutes, maybe a bit more, and in the end, after a dim, moist scramble that could raise a blush in her memory for years, body against body rattling anxiously out of fear of being discovered, Adele suddenly began to cry. Holy shee-it, Billy muttered, what did I do, did I hurt you? Without answering, Adele made a terrific effort to pull herself together. She was furious at her tears. Why was she crying? She wasn't unhappy, far from it. Reaching up, she kissed Billy primly on the cheek, embarrassed at last, then went home next door to try to figure it out. She would work at it in the bathtub. It was where she figured everything out.

Sitting there a few minutes later in the steaming aseptic green water, mindlessly curling her pubic shoots between her fingers, Adele thought about Billy Brent, about her parents downstairs, Max in his easy chair reading practically every word in the *Evening Sun*, Florence irritable in the kitchen at her own incompetence, probably burning the dinner again, about herself. But not for long. Adele never wasted time or motion. Her thoughts always took her right where she wanted to go. She had the true homing instinct. Her conclusion this evening was that she couldn't wait to see Billy Brent again, despite the tears. The tears, she decided, were a mere nervous effect of the episode, not so unusual where so much excitement was concerned, and irrelevant and contradictory, as well, like her parents in most matters. There was a long winter ahead, that was what counted, and she and Billy would have to make sure to mark off some of the cold dark hours in the late afternoons for themselves. They would use the Kates's garage. It stood there, airless and rusty, waiting for them. At the idea, at the intense, seductive images that rose in her mind, she lay back against the cold bathtub tile and cupped her left breast in the palm of her right hand, without thinking, like a voluptuous Eve she had once seen in an old German etching at the Walters Art Gallery (in the enforced company of Aunt Clara) and began, in a slowturning, dreamy, wholly abstract way, to think about Leopold Vogel, her future brother. In this way, she got rid of the afternoon.

Leopold was undoubtedly on everyone's mind these days. Florence was anxious about his arrival, so anxious that her migraines, which dated from Adele's birth, had begun to recur; Max, whatever he told the social worker, was not at all sure that he was ready, much less fit, to father a son; and Adele wondered silently at it all, mostly in vague generalities. An unknown young boy named Leopold Vogel would soon arrive. That was clear. That was real. Hitler would finally let him out of Germany, into exile. This idea was expressed in Adele's mind in various pinwheeling images in which swastikas and Nazi banners predominated. But beyond that it was hard to know how to consider the question of Leopold Vogel in any specific way. He was there in Hesse, across the Atlantic, she was here in Baltimore, Maryland; and for Adele Gordon reality was where she happened to be. However adventurous she was in a certain sense, however smart and cheeky in class at Forest Park High, it is possible that Adele Gordon would not have been able to find Germany, much less Frankfurt-am-Main, on an unmarked map of Europe. This ignorance of certain basic geographical facts was shared with almost everyone in her school, where few students seemed to know anything for sure; Billy Brent, among them, had never even heard of Frankfurt, Germany. For Adele Gordon, reclining so pleasantly in the bath after her great adventure (while planning her next), her left breast still resting lightly in her right hand, Leopold Vogel came from a tyrannized nowhere, haphazardly stuck in another country across the sea, and was still without a recognizable human face, without recognizable human features of any kind.

Within a week a little sleet fell on both continents, and city streets in Hesse and Maryland were thinly iced by morning. Fall was over. By then Florence Gordon was busy getting a small upstairs bedroom ready for Leopold, the one Max had indicated on the floor plans he had drawn up for the authorities. Adele helped her mother, by special invitation, enjoying the passing authority that Florence doled out to her piecemeal. It seemed a good chance for Adele to learn how to do something real. She was allowed to buy the curtains herself, she measured the single window for venetian blinds, picked out a shag rug downtown at the May Co. Max also made suggestions, speaking up helped to generate confidence in himself, he even found time to buy a set of pirate bookends for Leopold, which he thought he himself would have liked to have had as a boy. All of that was ordinary enough and the Gordons, all of them, enjoyed it. It passed the time, that was what mattered, it killed the lethargy that came with waiting. The furniture they chose was perhaps too coy, too self-conscious—the headboard of Leopold's new bed showed a rain-soaked little boy holding an umbrella blown inside out by the wind—but it was cheerful. Those were Florence's orders. Everything had to be bright and positive. That was how foster-children knew they were welcome, she said, how Leopold would begin to learn how to forget the past. Max's sister, Clara, chipped in by sending over a desk made by some Indians in New Mexico, which she had ordered by mail, and her husband, Lester, kindly came up with a bright and positive desk lamp on his own.

Soon there were promising signs of quick action. Final character references were suddenly asked of the Gordons, and in Frankfurt the Vogels had to sign an additional affidavit for the release of their son. A friend of Clara's and Lester's got Mayor Jackson to write a fulsome letter to the Jewish Welfare Agency about the Gordons, whom he had never met; it was hand-delivered from City Hall to the agency's office to underline the urgency of the matter. Then Leopold's dental file arrived unexpectedly from Frankfurt at the agency offices and was forwarded uptown to the Gordons, where no one knew what to make of it. It was a flurry, but it was serious; and it continued. At last, as Leopold's room on Fairfax Road slowly filled with cheerful Americana and Max Gordon's jewelry store downtown moved towards the Christmas sales season, Elsa Vogel and Florence Gordon began to write to each other. This obvious idea was suggested to them by their social workers, by careful prearrangement, and they welcomed it; it was such an easy way to exchange essential information and probe the future; perhaps, as both women saw it, even control it.

"Dear Friend," [began one letter from Frau Vogel]: "My husband and I tell each other a thousand times a day how blessed we are and what sacrifices it is for you and your own kind husband to offer to take over the direction of our son's life at this time. We thank you to eternity. In the midst of such madness everywhere (I am sure you will agree even if you are so far away), it is a Godsend to us and to him as we prepare his leave-taking. We his parents say please do not be shy with him about household work. He is a good boy, but he must learn his duties, like all children. He must come to understand, at his age, that manual labor and hard work do not harm his dignity . . ."

To which Florence Gordon answered: "Dear Frau Vogel, May I call you Elsa? It seems to me the natural way, only proper given our situation. And you must feel free to call me Florence, as my friends do here in Baltimore. That will be much more comfortable, don't you agree? We must not stand on ceremony with each other. I beg you not to worry, my dear. We will do all we can for your child. We are prepared and ready for the task. Everyone knows that my husband is very fond of children. He dotes, in fact. And my daughter is very excited at Leopold's arrival. She is beside herself with suspense. I am sure they will become like brother and sister, real pals, as we say here in America. She is a good girl and a respectful daughter. With God's help, all will turn out all right. Your son will be kept as though he were our own child. Be assured of that. You must have the utmost confidence in my husband and myself. Life now demands such solutions from all of us . . ."

These sentiments—the last of which was lifted from Max Gordon by his wife, who felt that it contained a stirring resonance, just the thing to express her present emotions—continued to be posted through the middle of November. They were accompanied by instructions from Frau Vogel, politely autocratic words set down in pale violet ink, about questions of diet, allergies, clothing, education, hobbies, sleep habits, even, God help her, Leopold's bowel movements ("I hardly know how to say this . . ." she wrote, introducing the subject); fervent exhortations from Florence Gordon for courage and faith, which did Florence more good than her correspondent; practical suggestions, borrowed from the JWA's vast sum of unhappy experience, for speeding Leopold's departure; improvised prayers; and an ongoing mutual flattery, fully conscious, that went a long way towards providing comfort for each woman.

They wrote to each other in English, but Elsa Vogel first composed her letters in German. Then she made a more or less literal translation, taking enormous pains over syntax and grammar. Every word was later checked and, if possible, colloquialized in his sharpest manner by her older son, Kurt, who needed regular exercise in English. Once he had finished working on the letter, pretending detachment at its contents, as though he were a teacher coolly marking a student's paper, Elsa went over it for the last time, making an inked change here and there. Then she let her husband read it. A family consensus, which excluded Leopold, followed. Should this suggestion be retained or dropped? Julius was for one thing, Elsa another. Was the tone acceptable here, was it perhaps too strong there? How would it read in America? Was the vocabulary clear, the words just right, did all those formidable sentences, which were so subtly deciding Leopold Vogel's destiny, make sense? Kurt was impatient. His parents had trouble making decisions. They were wasting too much time, he scolded. Life couldn't wait forever. Finally Elsa Vogel carefully folded the sheets of paper in three, paying attention to neatness and appearance, as though the folds contained a piece of her son to be displayed to Florence Gordon, a breathing organic part of Leopold Vogel, clean and precise, and slipped the sheets into an envelope. Then Leopold himself went out to mail the letter, without, however, being allowed to read it. He was the hero of every line, he lay gently cradled inside each word of the exchange, but all he ever saw of the correspondence in his walk to the mailbox was the curious American name and address, written in his mother's fretful hand, that filled most of the space on the envelope beneath the stamp showing Hitler armored for spiritual battle.

Occasionally, powerful feelings broke through the domesticated language. "Very sensitive young people," Elsa Vogel wrote at one point, with the authority of her nightmares, and Leopold's as well, "get upset beyond their years by such troubles. They seem to know, without precisely knowing, what is happening around them. They develop, as we say, the habit of disaster. I see it everywhere. To prevent this, they must be watched, guarded, protected. You will forgive me, will you not, if I say it in German, too, [this was Kurt's idea] for unforgettable emphasis." And there followed the three foreign words, *beobachted, begleited, bewacht,* underlined on the page in violet ink, the strangeness and thudding force of each verb coming at Florence Gordon with the urgency of a command that must not be disputed. She thought she had discovered in them the real sound of Elsa Vogel's voice, the solid, four-square rumble of Leopold's German mother. That voice, she found, carried heavy Teutonic echoes that could span continents without resistance, it had a transmittable energy, and Florence Gordon, sitting uneasily in her American rowhouse on Fairfax Road, in Baltimore, Maryland, pondering the letter in her hand, impatient with delays and government obstacles, intolerant of visas, German youth exits, official dispensations, the whole prickly apparatus of international escape and flight, would not easily forget it.

And then later, as the first timid snows began to fall, again from Elsa Vogel: "Conflicts are inevitable. Of course we are resigned to that and we hope you are too. He will have to learn how to adjust. Perhaps—please, without presumption—you will all have to learn together [another Kurt idea]. Leopold must not be allowed to think that all this is his *due* . . ."

But by then Leopold was no longer Leopold. He had grown tired of the name. It had become a burden through all the interviews and heavy paperwork of the past year. It seemed to him to belong to someone else now, someone ordinary, an everyday Hessian of the kind that chose not to speak to him on the streets of his city: an old classmate, for instance. With his thoughts turned daily to the West, to the mysterious Gordons and their unimaginable city, his name had begun to reflect something of the threatening, agitated heart of Central Europe that already seemed foreign to him; and something, too, of Frankfurt itself, containing all the German things he could no longer have, that were now forbidden. Frankfurt was where his name belonged, Central Europe was its home. He would leave it there, give up Leopold Manfred Vogel, become Manfred Vogel. Manfred stood alone, he told himself; and he would be alone. Manfred sounded manly; and he would be manly. Also contemporary; contemporary was important. Contemporary meant modern and modern meant America.

He raised the question after dinner one night, over cheese, fruit, and sweet Turkish coffee, for which his father had a taste. His brother faced him across the table, his mother was at his left, dabbing at her mouth with the corner of a damask napkin. She wore no lipstick. At the head of the table, his father worked clumsily at slicing a thick Muenster with his left hand, while keeping his right, which was missing the middle finger, hidden in his lap; the finger had been lost at sea in a Baltic naval battle in 1917. Leopold decided to take the long way round. He began by talking about school.

"Joachim Adler," he said, breaking the long silence at the table, "failed a history test again yesterday." He glanced around the room as he spoke. They might be listening, they might not. "One more and he'll be out," he continued. "That's what Herr Levi says, but that's what Herr Levi always says. They can't throw Joachim Adler out. He's got no place to go. It'll be too embarrassing. What real school would take him in, the son of a kosher butcher?"

"He's already in a real school and so are you," Julius Vogel said.

"You know what I mean. Real German school. Anyway, that's not the point. The point is that when Herr Levi began to threaten him, Joachim took out one of his old man's rusty knives, one of those little Swiss things with a dozen blades, and began to carve on his desk, very slowly and very deep, to drive Herr Levi crazy, those immortal words from *Mein Kampf*, and I quote: "It is not for this that I have waged a fourteen-year struggle."

Elsa Vogel hissed behind her napkin. "Don't be so smart," she said. "Keep those things to yourself."

"I do. I'm a walking tomb. But it really happened. It's there on his desk, carved in wood forever. And Herr Levi went crazy when he saw it. He always goes crazy, over anything." He folded his arms, waiting for their response.

"Why don't you have some cheese?" his father suggested, as though he hadn't heard a word. "The government nutritionists tell us it has high protein value and will provide all the vitamin A you need for the next twenty-four hours. Also, it happens to be delicious. I love Muenster, since I was a boy." He passed around the plate, filled now with thin crumbly slices. "Come on," he urged his sons, trying to avoid the discussion about Joachim Adler and Herr Levi, which pained him.

"They'll be carving that Brownshirt shit on your ass if you're not careful," Kurt said to his brother, as he peeled an apple. "You're just what they're looking for, just the right type, like your friend Joachim Adler, big fat sassy mouth and all."

"Don't leave the apple skin on the tablecloth," Elsa Vogel said. "Use your plate."

As his wife spoke, Julius Vogel turned to his older son, taking his time about it. He would not be hurried, his posture said, he would not let anyone rush him. That was what his authority had finally come to, a couple of stiff, slow-moving gestures tentatively displayed at the head of the table. "You may be twenty years old," he stated, concentrating on his dignity, "but in this house you still have your mother and father to consider. I don't care how you speak to your brother when you are alone with him, or how he speaks to you, for that matter, but here at this table, at *our* table, sitting with your mother and father, dining on their bounty, as it once was called, you are still a guest of the household and you will show us respect."

"Indeed," Kurt muttered, looking away.

"*Indeed*," his father said. He began to play with the bread crumbs in front of him, moving them around as though they were chess pieces. "Here, as I say," Julius went on, despite himself, hoping that he could end the exchange with a single sweeping statement, "here you are still a guest. Good manners insist on it. Good Vogel manners. You won't speak that way at my table, as long as it remains my table."

Kurt Vogel hung his head in mock-shame. It was a favorite pose in front of his family. He even managed to blush a little. While Leopold thought of America, Kurt thought of England. He was halfway through the visa process and on his best behavior, trying to keep himself as neutral as possible in all situations. He no longer talked politics on the outside, and there would be no serious disagreements among the Vogels at home, if he could help it. "Yes, Papa," he said, in a voice full of false humility. "It's that I worry about Leo. I don't want him exposed to another attack. Once is enough, yes?"

"It's Joachim Adler who's in trouble, not me," Leopold said.

"Remember, the walls have ears," Elsa Vogel stated, pointing vaguely to the right and left.

"She gave me a piece of candy again," Leopold said.

Julius Vogel and his wife exchanged a sharp look. "When?" Julius asked.

"Yesterday."

"Where?"

"I met her on the stairway. I was going down, she was coming up. She gave me a piece of fudge. It had lint on it." He did an imitation then of a tiny old woman, bent nearly in two, with palsied hands. His shoulders hunched over, his mouth folded in. His hands, which he shaped like claws, trembled in front of him, offering a piece of invisible candy to the vacant air.

"Bravo," Kurt said, applauding. "It's *die Alte* to the bone."

Even Elsa Vogel had to smile at the performance.

"Why is she always knocking at our door, day and night?" Leopold asked.

"Because she is lonely," Elsa Vogel said. "Because she has no one to talk to."

"Neither has Herr Karton downstairs and he doesn't knock at our door."

"Herr Karton has a daughter in Pomerania and two grandchildren," Elsa Vogel said.

"We're avoiding the issue again," Kurt said.

"Don't worry so much about the issue," Julius answered, with a bleak look.

"She's blackmailing the Vogel family."

"I said the walls can hear," Elsa Vogel warned again.

"Is she blackmailing us?" Leopold asked. "Is that true?"

"Papa pays *die Alte* thirty marks a month," Kurt said.

"For what?"

"So she won't say anything damaging to the authorities, that we are hoarding Muenster or that you keep a map of the USA on your wall."

"Is that true?" Leopold asked again, facing his father.

"Everything is true today," Julius Vogel said loftily, over his son's head.

"What does that mean?" Leopold asked.

"He only means that everything has come true," Kurt said to his brother. "As it has."

There was a silence then that lasted a full minute. Certain subjects in the Vogel household were unbearable to pursue.

"Well," Leopold finally said, sitting up impatiently at his place. "Can anyone in this family tell me why Frau Göring chose to leave the church on her honeymoon?"

"No," Kurt said. "Why did Frau Göring choose to leave the church on her honeymoon?"

"Because she had lost faith in the resurrection of the flesh," Leopold said, barking a laugh at his brother.

"Don't talk so fast," his mother complained, covering her ears with her hands. "It makes me dizzy."

"Leo," Julius Vogel objected dutifully, but this time he laughed at the joke with his sons.

At the sound of his name, a tic under Leopold's eye flashed a warning. He began to poke at an old patch of eczema on the inside of his elbow. He swallowed some air and let it out with a thumping sound that was not permitted at the Vogel table. His mother gave him a disapproving look. "And you will do that when you go to America?" she asked.

"Do what?"

"Belch like a pig," Kurt said.

"I didn't belch."

"And tell filthy jokes, too?"

"Oh, filthy jokes." Leopold looked pleased with himself.

"Leave the boy alone," Julius Vogel said. "You can be sure they belch in America, too."

Leopold plunged in while he sensed an ally. "I want to be called Manfred," he said, head bowed in front of his family. "I want to be called Manfred when I go to America."

Kurt raised his eyebrows. "A brave new name for a brave new world," he said.

"Manfred," Leopold repeated, rushing himself a little, even though there was no resistance for the moment. "I don't want to be Leopold." He looked up.

"And what's wrong with Leopold?" Julius Vogel asked.

"It's old-fashioned. It's a German name. I don't want to have an old-fashioned German name in America."

"Manfred is a German name," Kurt pointed out. "And you're a German boy. Or have you forgotten?"

Leopold threw his brother a fierce look, biting his lower lip to stay in control.

"So you don't want to have a German name in America," Julius Vogel said, in a thoughtful singsong. "Are you ashamed to have a German name? Is that it? Come, you can be honest. I have a German name, Kurt has a German name, after all, we've been Germans for hundreds of years. You've forgotten that already, you're in such a hurry to run off. There were Vogels in the Rhineland in the seventeenth century. Your Darmstadt grandfather fought in the Hessian infantry in '70. He was only seventeen, a volunteer. He went all the way to Paris, the first Vogel in Paris. Jenny has a letter from him describing Yom Kippur services before the great offensive outside Metz."

"You've told me a hundred times."

"Once more won't hurt," Julius said sharply. But he instantly pulled back. He was talking too much. He couldn't help it. The sound of his own voice in his own home helped him to believe in himself. His nose began to twitch now, as though it sniffed an ambush ahead. "I say again," he went on, carefully now, "that your grandfather was a German. He was one of the first occupying troops in Paris. He slept in the Champs de Mars, near Napoleon's tomb. He bivouacked along the Seine. He was awarded the Iron Cross . . ."

"Second class," Leopold interrupted.

"The Iron Cross, I say again," Julius went on, raising his voice. "Aunt Jenny has it in her bank vault. It's for you or Kurt someday. And your grandfather's father was a German, remember that, and his father and *his* father, and back, back."

"But I'm not," Leopold said firmly. "I'm *not*," he insisted, at his father's stubborn look. "If you don't believe me, just ask *die Alte*. She'll tell you what I am and so will Frau Göring's husband."

"I don't need Frau Göring's husband to tell me anything," Julius snapped, but a moment later, facing them all, he looked and sounded as though his mind had suddenly flown off elsewhere. "What are we arguing about?" he asked vaguely. "No one calls you Leopold anyway. You're Leo. You've always been Leo."

"Not anymore." Leopold decided then to resolve the question; his sense of timing, precisely honed by years of practice at this dinner table, urged him on. "It's done," he said. "I don't want to talk about it again. It's not a matter for discussion. And—" turning to face his parents, mother first, then father—"I don't care what the official papers say, don't try to use them against me, they're just official papers, they're for strangers. I want to be Manfred, I am Manfred, and I want you to tell the lady in Baltimore—" he never referred to Florence by name—"I want you to tell her that I want it like that."

"But Leo," his mother began, reaching out to him.

"Manfred," he shouted.

"Gott," Elsa Vogel said. She put a hand to her brow.

"Mama," Leopold said. "Don't look like that. Please."

"Manfred," Julius said under his breath, trying out the name experimentally.

"Well, I like it," Kurt said, breaking in as though the final decision lay in his hands. "New lives need new names."

"Nobody asked you," his father said.

"I still like it. And as for new lives . . ."

"What do you know about new lives?"

"As much as anybody."

"Is that so?"

"As a matter of fact, it is. I'm going to have a new life myself. Leo's not the only gypsy around here."

"What does that mean?"

"I'm going to London."

"To see the queen?"

"You'd better take me seriously. I'm going to London, but under my real name. Kurt Vogel, *echt Deutsch*. My visa is being prepared now. All the papers are in the right hands. It's a surprise, yes?" He shrugged doubtfully as he examined their faces, which gave nothing back at his news. "You know I have good references from the university, important names, everything that's needed. Professor Schleiter will do anything for me while he can, he's out on a limb already, he's pushing for Cambridge . . ."

Elsa Vogel's cup rattled in her saucer. She put it down carefully, taking her time about it. Where had all this come from? Why was she so unprepared? It was so obvious, the logic of it so powerful. How could she have not expected it? Cambridge, she thought; Professor Schleiter with his Prussian energy and cosmopolitan enthusiasms. He had been Kurt's favorite at the university and Kurt had been one of his. It was a simple equation. All it came to was that she would lose another son.

"Well," Kurt said, his voice full of strained enthusiasm. "Doesn't anybody care?"

"But Kurt. . ." Julius Vogel began. But Kurt—what? Julius didn't know. He found himself making a placating gesture with his right hand, characteristically defensive, palm out in immemorial style, the stump of his middle finger, netted by wrinkles that were like fine pink hairs, shining tautly as he held it up in the light. He would not resist. Kurt would go, too. "I always wanted to see England," Julius finally said. "I always wanted to walk along Piccadilly. So you beat me to it. You win. But the English are not so interested in young Germans who want to become English. Isn't that so?"

"I will make them interested."

"Ah, you will make them interested."

"Make them," Kurt repeated. "And so will Professor Schleiter."

"Come to America," Leopold broke in. "Come with me."

"If I had to wait for a U.S. visa . . ." Kurt said, with a shrug. He took out a small cigar and lit it with a kitchen match that he pulled from his shirt pocket. "If I have to wait," he went on, puffing blue smoke over their heads, "I won't get out. I'm doing what I can for myself, all in the spirit of self-interest. Beats there a heart anywhere that is not fed by self-interest? Shakespeare? Goethe? One of them said it, it sounds like the Weimar ghost to me. Anyway, it's true and we all understand the truth at this table. Right, Papa? Schleiter encourages me. I'll be in England in six months."

"In six months this will all be over," Julius Vogel said. "I will have the business back, what's ours will be ours. Even this winter, things are different, change is everywhere."

"Olympic propaganda shit," Kurt said, making one word out of it. "You know my secret now," he added. "Now pretend you didn't hear. Not a word to anyone. The walls have ears."

Dinner wound down. The apple core in front of Kurt had already turned brown. A slice of splintered cheese lay on the platter in the center of the table. Elsa Vogel's eyes, watery from the threat of cataracts, wandered haplessly from one son to the other. She raised her napkin to her lips again, as though to hide something, but there was nothing to hide, there were no words in her head. Julius Vogel stared mysteriously into space, stirring a cup of cold coffee.

"One more," Leopold said.

"One more what?" his brother asked.

"I just heard it in chemistry lab. Even Herr Levi had to laugh. It is only now," Leopold said, in a solemn voice, "that we can understand the true meaning of Captain Roehm's message to the boys of Germany."

"And what is this true meaning?"

"The true meaning of Captain Roehm's message to the boys of Germany is that out of every Hitler youth a storm trooper will emerge."

"Oh, God," Kurt said, grimacing, while both his parents gave a kind of snort. "You'll get put away for that kind of thing. Where do you pick up that stuff? You'll go to jail or worse for it."

But Leopold knew that he had forced them to pay attention at last. They were talking to each other tonight because of him. They had spoken angry and acerbic words over dinner. Kurt had shown his teeth; his father had answered; his mother actually laughed at something he had said. What would they do when he deserted them for America? How could they go on without him, where would they find their happiness?

A calm sea of starched white Vogel tablecloth stretched in front of Leopold. His own damask napkin lay pyramided alongside the Vogel fruit bowl. Outside, the early winter rains had started again; the water froze as it fell, turning into sleet before it hit the pavement. Leopold looked around the dining room without expression, examining it in the sugary pink light of his mother's chandelier. The light blurred; his eyes blinked with fatigue. The chandelier had a long mournful history, a sad narrative all its own, like one of the lost tribes of Judah. He knew it by heart. It came from the Manns, his mother's side, along with the Bohemian crystal and the Dutch porcelain, ran crookedly for decades from Amsterdam to Kassel, Kassel to Darmstadt, back to Kassel for a generation, and finally to Frankfurt and the Mendelssohnstrasse. It had survived four wars, two revolutions, and one local rebellion. It was a European commonplace, like the crystal, the porcelain, the silver from Spain, and Julius's library of classic German literature, which rested authoritatively on the Vogels' shelves, sturdily bound in crimson tooled leather, volume after volume of multiple inherited worlds imprinted on thin silky pages.

The pink light wavered above the table. His mother's eyes seemed to liquefy. Kurt pushed his chair back, preparing to get up. There was a soft flurry of noise from his father as Julius cleared his throat. He moved slowly, almost weightlessly, a small man with small feet, thin fine nose, thin crinkly hair. Still Leopold sat there sphinxlike, a Vogel like the rest, like his father and brother, thin and fine-boned, feeling the pleasant, substantial weight of his new name rise slowly through his ribcage. Quietly, he tested it.

Man-fred.

Chapter Three

He was told to get ready. In a few days he would leave for America from Bremerhaven. His mother waved the steamship tickets in his face. "You see?" she cried in a furious voice. "This time it is not a false alarm." There had been an aborted alert from the agency a few weeks before, which had temporarily exhausted their nervous energy and left Manfred chewing his nails. Elsa waved the tickets in the air again, reaching at the same time for his suitcase under the bed. "Help me," she ordered. "Don't stand there like a damn fool." The real excitement had begun, a powerful strand of Vogel hysteria had been let loose. But first, a letter arrived from Adele Gordon, for Manfred's surprised eyes only. *Personal*, it said on the envelope, underlined twice. It was written on large looseleaf sheets, with three reinforced holes in each lefthand margin. "Dear Leopold," it began, in all innocence. At the top of the page, on the righthand side, was a cartoon drawing of a smiling face, all fat circles, surrounded by a great cloud of frizzy hair. "This is NOT me," the caption read.

"Dear Leopold," he began again, glancing at the drawing a second time. Dear Leopold. This stupid Yankee Doodle doesn't even know my name, he mumbled to himself, and plunged on. "My mother thinks it would be a good idea if we write each other. Pen pals, that stuff. She's probably right. Sometimes she is, sometimes she isn't. When she isn't, she still thinks she is. (With me, facts are facts.) I don't let her get away with anything, that's the rule, but I ALWAYS SMILE when we're going at it and that's my advice to you. Keep smiling. (She's a sucker for a smile, who isn't?) End of advice.

"Anyway, we're going to be close to each other all right, like it or not. Personally, I think I like it. One kid in my experience takes up a lot of room in a house, two kids maybe less. Have you ever noticed that? In English, that is what is called a paradox (I think) but you are fluent in the language, or so they tell me (God, I hope so), so you probably know all that. I'm not fluent in anything. I can say venividivici, jaime, commentallezvousilfaitfroid, that stuff, then I'm dead. The English language around here, so you're not disappointed when you hear it, is not exactly Ronald Colman purring away to a lot of pretty ladies in hoopskirts, it's more a lot of uh-uhs, y'knows, see yas, y'all, that kind of thing. People can yak for an hour like that. One syllable at a time. Sounds not words. Sometimes Baltimore talk seems like a bunch of cats wailing. Seems to me there's a lot of whining and meowing. (Do cats meow in German? Joke!) But it's English all right. We all understand it. We manage, even though I think English is full of mean tricks and I can prove it. Things like tough, thought, though, trough. Who made those up? How did they get there? And there are others just as mean. Bin, been, I, eye, aye, hi, hie, high. And heigh, as in ho. The language is full of 'em. You'll see what I mean when you get here—here, ear, and early, much and touch—you'll get the hang of it, but for God's sake put it out of your mind for now, I didn't even know I was going to write this, IT JUST CAME OUT, IT'S NOT A SERIOUS THING.

"Your bedroom will be right next to mine. Lucky kid. The wall is about as thick as a piece of toilet paper so I can hear everything, EVERYTHING, and we will be using the same bathroom, too, there's only one, along with my mother and father. I like to sit there in the tub, but everyone else is very fast. My mother, whoosh, my father, you can hardly even tell that he uses the bathroom. They tell me it comes from growing up in a big family, which they both did, five kids in one, four in the other, where everybody has to make way for everybody else. Well, maybe. I don't know. I don't think I make way so easily myself, it's a fault.

"I am practically seventeen and in the twelfth grade, very good marks but lazy. My parents complain, even though I am ahead of myself. The school is OK, girls and boys at one time, we call it coed. There is a lot of talk now about whether you should go to a boys and girls or all boys. If you go to an all boys, you'll have to travel on a streetcar to get there. You use a token that costs a nickel. Boys and girls, coed is better. Boys and girls is better for lots of reasons. (You're not afraid of girls, are you?) You tell them you want coed when you get here.

"I am destined for college, if we can afford it when the time comes. Destined, what a beautiful word. BEAUTIFUL. College is where you get blessed with all the good things in life for four years and come out gold all over. If we can't afford a real college, there's always the University of Maryland, a couple of miles down Route 40. It's a discount operation called College Park. If you live in Maryland, it costs about fifty cents a semester. Food and a bed is extra. You live in a dormitory. Girls in one, boys in another. I wouldn't mind College Park, I guess. You come out copper instead of gold is all. Affording it is one of the questions around here. Around Baltimore, I mean. Can you afford this, can you afford that? The money in this house comes from Daddy's jewelry store downtown. There's enough, I guess. Daddy goes to museums sometimes, about once a year, and draws dream plans for houses on graph paper. Or he used to before he got too busy with the store. All the Gordons are like that. A little artsy. You know?

"Actually, Daddy's adorable. His hair is prematurely white. I mean really white, the kind that's on the other side of snow, and he has dark coloring that lucky people with white hair sometimes have. Italians, people like that. You know? I mean he looks sunburned all the time, in the winter too, and his hair goes back in waves, and you just keep wanting to touch it. I mean I do. Lovely, oh lovely. Capicecomprenezvous? You'll like him. You better. (Ha-Ha, but I'm serious too.) He has his little habits, his little things. Who doesn't? My mother runs everything around here. She thinks. That seems OK with my father. You can be sure he doesn't complain to me. But I'm not so sure how he feels. You never know about a lot of things. When they fight I used to hide in the closet.

"We have some books in the house. I thought you'd be interested in that. *The Book of Knowledge*, *The Life of Abie Lincoln*, things like that, a big Philco in the living room next to Daddy's chair, and a new portable for my mother. She keeps it in the kitchen. My mother likes to listen to the radio when she cooks. She says it takes her mind off the chopping and the slicing and watching stuff from boiling over. And I can believe it when we sit down to supper every night. I mean when you take a mouthful you can tell her mind was on something else in the kitchen. "Stella Dallas," "Our Gal Sunday," one of those. You know it wasn't on the food, that's for sure.

"By the way, can you read my handwriting? Everybody complains about my handwriting. Nobody can read it. I can, it seems clear as peaches to me. But in school. Oh, in school. Can you read it? Suppose you can't? How will you know what I'm saying?"

Manfred paused a moment, looked up, batted his eyes. It had taken him fifteen minutes of agitated strain and constant backtracking to get this far. Could he read her handwriting? Well, yes and no. The script was huge and swirling; no one in Frankfurt wrote like that. Little circles mooned above the thick *i*'s and the *t*'s were crossed with long bars that flew across the page, *w*'s and *m*'s were arching worlds to themselves, and letters that went below the line, all those great dripping loops, in most cases went below two lines. Punctuation was scarce, or so Manfred judged, spelling careless, shorthand commonplace. It was not like school English, no, it was certainly not like Herr Levi's careful constructions. But Manfred liked it; it tugged at him with its alarms and threats, its fervor. At the very moment he was reading it, a kind of unembarrassed presence, a full, wheeling amplitude, came right off the page on its own. That part was irresistible. He looked at the cartoon face again, way back on the first sheet of paper, tried out a little smile on himself, and went on.

"My Aunt Clara," Adele wrote, "can't wait to get her hands on you. She already has her hands on me, trying to turn me into a debutante or something before my time. I can never be a debutante anyway. It's not for Jews, you should know that. (It's not for boys, either.) She wants me to be cultured and she will want you to be cultured. Are you cultured already? I hate the word. I can't even say it out loud, even though I am very cultured (as well as vulgar). Can that be? Does one go with the other? It doesn't with people like Aunt Clara, but it does with Adele Gordon, I guess. Paradoxes, paradoxes. I love 'em. Also, I will have lots of friends ready for you. Good American types, VERY VULGAR! Billy Brent down the street can't wait to get his hands on you (I'll protect you, don't worry). Laura Piscitelli, too, in the house on the corner, sometimes she gets on my nerves, she's so weird sometimes, she's Catholic, her whole family is, and Jojo McAllister, really John Joseph, who is so sweet. He never complains, he grins all the time, even when they keep him back in school. There are about sixty other kids on the block, two to a house, mostly dopes, and there are forty-three houses, each one stuck to the other, we call them rowhouses, with the same little front lawn, same little front porch, ditto backyard and tin garage, and an alleyway, too. Actually, if your eyesight is good and you know how to look you can tell that every house is different, not the same at all, even if your eyes are half-closed when you go past. You can tell there's a special kind of maniac living in each one. This maniac likes begonias, that one likes snowballs, they're all over the place, another shrubs, some have green-striped awnings on the porch, some have rickety old roll-up screens. Mr. Piscitelli is crazy for stone urns with little fir trees in them, and a lot of maniacs have zilch. That's only on the outside. Inside, it's worse. Maniacs everywhere. Canary maniacs, tropical fish, stamps, and nudies, too. You wouldn't believe it, what goes on inside. My daddy is a sycamore maniac. That's a tree. We have one that came with the house and drops smelly old balls on your head around this time of year. (NOTHING VULGAR INTENDED.)

"Anyway, I await Your Highness. Baltimore isn't bad. It really isn't. Hot in the summer, cold in the winter. What can I say? Great sleigh riding, great hill, right around the corner. You have a nice room, there's a little porch off it where you can sit in the summer and fry in the sun, like in the electric chair. I'm kidding. It's really nice. One window and a door to the porch with a screen on it for the flies. A new bed and a real Indian desk. I bought the curtains and the rug for it. My mother hates the rug. The closet is small, but OK (if you're a midget) . . . Listen, I'm tired. I'm not used to writing letters. I never write letters. You can believe half of what I say. Which half is for me to know, you to find out. If you're smart. I'm kidding. I'm always kidding. Around here, they tell me you're supposed to be smart. That's what they say, high IQ. Better that than the other thing . . . Looking forward now," and there followed Adele's great wandering signature, dominated by the fattest *A* Manfred had ever seen. At the bottom of the page, there was also another cartoon face, this one with eyes closed and mouth opened in a yawn. The cloud of frizzy hair was in place. "Time for beddy-by," the caption read. "See ya soon. Tell your family I'm not so dumb."

He told his family nothing, although they asked about it often enough. He kept it all to himself, refused to share a word. "And what does the beautiful young American fraulein have to say?" Kurt teased in his most sardonic tones. "Be sure to bring galoshes and clean underwear?" "That's right," his mother added, becoming suddenly aggressive, as she sometimes did. "Don't be stingy. Tell us." But he would not be pushed. He waved them off as though they were a cluster of beggars. Kurt poked him in the ribs. "Come on," he said. "We're all one sack of potatoes." "Quit bothering me," Manfred said, turning away. He could be as stubborn as any of them. Moving casually, he took a few steps across the room in order to separate himself from his family.

In the background, his father quietly examined him out of the corner of his eye, while seeming to be checking his own mail. "Leave him alone," Julius Vogel finally said to Kurt and Elsa. "It's his private property." "You heard Papa," Manfred said, from his corner. "It's addressed to me. Personal. See?" And he held the thick envelope over their heads for a moment, tantalizing them all.

Days passed. A call came from the agency. The departure was set. The steamship tickets had to be validated. There would be a final briefing for Manfred on Friday morning. He would be joined by four other Hessian boys on the trip. Two were going to Cleveland, Ohio, one to Louisville, Kentucky, one to Scranton, Pennsylvania. They would share a third class cabin on the *Europa*; the fare was a hundred-and-one dollars. Julius reconfirmed the instructions himself, making the trip to the agency office on Konigswarterstrasse for last-minute counsel and reassurance. The family could not face a postponement again, he told them. It would be too much of a strain, none of them could take it. His wife's heightened sensibility, he explained, so touchy these days, so unpredictable, left her especially vulnerable to unexpected nervous attacks from time to time. And he couldn't speak for Manfred, he might collapse, he might come apart, anything could happen. Of course, the social worker said, staring dispiritedly at the papers on her desk, we know exactly what it is, you're not alone. But that wasn't all. There were other problems Julius didn't mention. He had discovered that he had begun to misplace important things, Manfred's identification papers and medical affidavits among them. It was a disconcerting new habit that seemed to be without explanation, and he immediately sensed that it might be permanent. His wife had soon found the papers in one of Julius's desk-drawers while cleaning the apartment, but the next day something else was gone, and the day after that, and Julius was now afraid that he might lose something of Manfred's that would be both crucial and irreplaceable. The journey had to begin. There were suddenly too many variables.

Meanwhile, Manfred's open suitcase sat airing on one of the windowsills facing the courtyard. Piles of laundered clothing lay on the bed. The tickets were in hand. His mother waved them angrily in his face every day. His eczema spread. How would he get to Bremerhaven, his parents asked. Could he go alone? Should they make the trip with him? "He's going to travel halfway around the world without us," Kurt reminded them, "and you're worried about a piffling little train trip?" The agency suggested that Kurt himself accompany his brother, particularly since there were no *kindertrains* scheduled at the moment. It didn't seem a bad idea. It had its psychological advantages, the social worker told Julius. It would save Julius and Elsa the excruciating anguish of the final separation, or, failing that, help to hide it from their son. Yes, yes, Julius agreed, coughing strenuously from sudden emotion. Yes, he repeated, he thought that made sense. Everyone should be spared the worst, especially his wife. The social worker, who knew everything about such matters, was pleased with Julius's response. He had been a good pupil all along, attentive and acquiescent, with a pleasantly fastidious manner, but he couldn't stop coughing now. Ah, she said sympathetically, offering him a glass of water, try to control yourself, Herr Vogel, this happens every day, please . . .

The future was now clearly in sight. It was visible as soon as Manfred opened his eyes in the morning. The tickets in his mother's hand, the piles of clean clothes in his room, the open suitcase, everything confirmed it. For the Vogels in Frankfurt, things would go on as they had. It was his parents' choice. To assure the future, they paid protection money every month to *die Alte*, the neighborhood informer; then they defended her at their own dinner table, as though she were a cherished ally. Manfred did not want to think about *die Alte* or the apartment off the Mendelssohnstrasse or his father's business or even Kurt's secret visa. He had already begun to live apart from his mother, father, and brother. Adele's letter, which he kept hidden in his pocket so that he could memorize it line by line when he was alone in his room, had finally severed him from the others; by now it had carried him halfway across the Atlantic. Who were these proprietary strangers who still surrounded him at every meal, asking impertinent personal questions? Manfred hardly recognized them from one day to the next; when they spoke to him, he hardly heard a word they were saying. Once, caught dreaming about American abstractions at the dinner table by Kurt, his eyes trancelike and remote, he actually had to take a moment to remind himself of his brother's name. Already, he was on his own.

That was how Manfred Vogel—with the unexpected help of the sister he had never seen—went about getting himself ready for the stupendous departure from Frankfurt.

Chapter Four

One by one, they filed into the Vogel apartment, majestic in their slowness, full of a certain false assurance that was all on the surface, prickly, self-absorbed and righteous, as though they still held the world's fate in their hands.

They sipped tea or sweet wine from the Moselle; ate cakes baked by Elsa; perspired fiercely even though an icy drizzle was falling outside; sighed without end and without knowing it; spoke about Manfred as though he was not in the room, or, perhaps, had suddenly gone deaf; felt morally superior while others were boasting about themselves; offered going-away presents; smiled rarely; ate more cakes, drank a second glass of wine; tried to hide their flatulence; avoided politics; toasted the émigré.

They were pretending to have a party.

The Vogel side arrived first, as they always did. They lived nearby, just around the corner and down the street, and had come on foot through the bad weather, complaining every step of the way. Bunions, corns, callouses, all tightened up in the rain and pinched. It was hard going that night on the streets of Frankfurt. There were three Vogels, bound tightly by conventional habit: Uncle Selig and Aunt Ernestine, his child-bride of thirty years, and Selig's and Julius's sister, Jenny. Jenny was the middle child, Selig the oldest. Julius therefore was their petted "baby." Jenny was also the family war-widow and still voluptuous in the old pre-1914 fashion, breasts heaved up like artillery earthworks, the rib cage itself caged by a formidable corset, stomach painfully flattened, buttocks at alert; her body had been at full military attention ever since she was a young woman, engaged to one of Kaiser Wilhelm's infantry corporals. In Jenny's bank vault lay her father's Iron Cross, second class, the souvenir from France. Wheezing heavily from the climb now, clutching at their hearts as they reached the top of the stairs, still complaining under their breath, but not without pleasure, the three paused for a moment on the landing, shook out their umbrellas, and stood them dripping in an old rusted stand. Julius was waiting for them in the open doorway.

"They're all right out here?" Jenny asked, when she finally got her breath back. She pointed at the umbrellas. "Nobody will take them?"

"And where do you think you are?" Julius answered testily. "In some beer-hall in the East End?" He forced a laugh, without humor; the truth was that he had been talking like that all day. A chronic irritability had suddenly taken hold of him.

Jenny entered the apartment ahead of Selig and Ernestine, pushing past Julius. "You know how these things are," she said, looking peevish. "Nothing is safe around here anymore." By "around here," Jenny meant the homeland. It was her unconscious code-phrase, invented spontaneously right after she heard the news of Manfred's beating near the Schillerplatz. Since that day, she had not been able to bring herself to say the name of her country aloud. It was either "around here" or "this place," with a bitter emphasis that varied with her mood. Once inside her brother's apartment, she moved slowly towards the dining room, casually checking the Bohemian crystal and Dutch china laid out with narcissistic care by Elsa, unconsciously counting each piece, another habit, before touching her sister-in-law's cheek with her own. "Take my raincoat, darling," she called to Manfred over Elsa's shoulder, "and hang it in the bathroom to dry."

Selig, Ernestine, and Jenny lived in the same building two hundred yards away. Jenny's apartment was one flight up from her brother and sister-in-law. That way, they could keep an eye on each other. If Manfred stuck his head out of his own living room window, he could just make out one of the steep gables of Jenny's flat tucked away beyond the Free Church spire. The gable was made out of yellow brick from Saxony and was framed beneath a red-tiled roof. At night, long after dusk, Manfred could sometimes see a reflection of misty streetlight wash softly upwards over the bricks and tile, a glimmering canary-and-pink aura that slowly changed color as the moon moved across the sky, before vanishing in a haze over the rooftops. Dimly shining like that, tinted romantically in the distance, it gave a sense of passing mystery to the widow Jenny's protected life, even to Selig's and Ernestine's, suggested unknown lovers and unspoken erotic adventures to Manfred, who liked the idea of both as far as Aunt Jenny was concerned; but that soon vanished in the darkness, like the moon. There was no mystery about his aunts and uncle when they visited him in his own apartment. They were what they were, they never changed. At the ring of the doorbell, trying to catch their breath in the habitual, desperate way, Uncle Selig sometimes turning his head aside to spit into his handkerchief, in came the same stout, short-legged, vein-clogged, proprietary relatives, some of whose timeworn genes—to Manfred's astonished disbelief whenever he thought about it—had to be exact twins of his own. How can that be, he asked himself over and over, standing in front of the mirror, examining himself; he resembled neither of them, they did not resemble their younger brother, Julius. But at night sometimes, at that remove, in that strange reflected light . . . Here, he wanted to shout over the rooftops, breaking in on their lives. I'm here. Me. Manfred V.—or Leopold, as it was then before he knew he was going to jilt them for America.

After a while, a fashionable quarter-of-an-hour or so, Uncle Selig, Aunt Ernestine, and the widow Jenny were followed by Elsa's clan: another trio, this one sleeker and somewhat surer of itself, a prideful self-possessed triangle made up of Uncle Oskar Mann and his wife, Lotte, and Oskar and Elsa's younger brother, Benno. Oskar and his wife, Lotte, were the family worldlings and so, for that matter, was Benno, in his own way. They entered the apartment noisily, at least Oskar and Lotte did, shaking off the rain like beautiful clipped poodles, shouting greetings to the Vogels with extrovert passion. It was how Oskar and Lotte entered first-class hotel lobbies all over Europe. Wherever they went, there was a minor commotion surrounding them—"the waltz of the moneybags," Kurt called it, inimitably (and enviously)—from which Benno and the rest stood quietly apart.

Oskar Mann, of course, was rich. He was rich in any terms. His money made him dear to many people who had never met him, but hoped to, and especially dear to some of his family, Elsa among them. Elsa couldn't resist the idea of Oskar's fortune; money gave Oskar a unique glamor in her eyes, it enhanced her family feelings, made her feel safe and powerful in her own right. Her brother's reputation as a stockbroker reached as far as Berlin, he was famous in financial centers all across the Third Reich, and beyond turning a profit as often as he had to without humiliating his competitors, he knew how to spread his money around, where to spend it, what to keep, how to stash some of it away in the right places. He also knew which men in powerful positions needed it. The knowledge of all this had always thrilled Elsa Vogel.

Like his money, like almost everything Oskar Mann dealt with, his wife was another ongoing asset. She was smart and quick, with a sharp, insistent intellectual edge that some people were afraid of; but it was not necessarily unkind. Lotte seemed to feel at home wherever they went, was comfortable in all her husband's worlds. She had made that her business, without complaint. And she brought with her an advanced degree in Romance Languages from Marburg, a practical help on their travels around the continent, as well as a badge of cultural authority that she was apt to flash on almost any family occasion. But in reality showing off embarrassed Lotte, and throughout her marriage she had always fought against the temptation. Everyone respected her for that. Elsa applauded it. Her sister-in-law was not a snob, never had been, and neither was her brother Oskar. Still, as they all had learned on one occasion or another, it seemed that it was hard not to show off sometimes.

Before they settled down for the evening, Uncle Oskar was already telling a long story about their Italian taxi-driver, who had just dropped them off downstairs, his voice, as he spoke, full of genuine wonder at the exceptional things that were always happening to him. "And then," he said, pacing the floor of the Vogel apartment, while everyone remained silent, "and then he had the nerve to say to me, and not only to say to me but expect me to believe, as Lotte and Benno will confirm, that Mussolini was his first cousin, on his mother's side, and that he only had to pass the word to Rome in order that . . ." Julius began to yawn. He had heard a thousand stories just like it from Oskar, many of them without a point, too. This one was standard stuff. But thank God for Lotte. "That's enough," she warned her husband, as Benno hung up their coats. She knew how to move in on cue. "Everyone is bored, as you can see. The endless charm of your life—and mine—is clearly resistible in this family. And you have it wrong, anyway, that's not what the dago claimed." She gave a good-humored laugh, her famous deep rolling chuckle, and turned to stroke Ernestine maternally on the cheek. She was fond of Ernestine, in her quick breathy way. They were both childless, Lotte from feminist principles, Ernestine from fear. Oskar shut up and took a glass of wine, not at all perturbed at the scolding.

This was an unusual gathering for the two families. Vogels and Manns rarely came together, almost never gave each other a second thought, except on the High Holidays when they broke the Yom Kippur fast in each other's company. For that purely social ritual, once a year, they became a unified clan. It was a little strange and a little forced, too. The custom had been set a few years after the war, while the widow Jenny was still in strict mourning for her dead husband, the Kaiser's corporal who had been lost somewhere behind the English lines in the confusion at the Somme and never found. It soon established itself as an annual event. Each year the families rotated the tradition among themselves, passing it from one household to another, including Benno's bachelor rooms when his turn came up. The ingredients never varied; the families made sure that the routine remained unchanged. That somehow seemed to be the essence of it: immutability, a fixed point, the cement of custom. Over the decades, it had come to represent to both the Manns and the Vogels a kind of primitive tribal emblem that might have descended from Sinai itself, it was so insistent, mark of an innocent repetitive pastime (which they nevertheless would have killed for rather than relinquish), blandly flavored each year by fruit juice, hard-boiled eggs, herrings from the north, tomatoes, cheese, and black bread. It was a menu developed over the centuries to cut the effect of the fast without causing too much discomfort; breaking fast had its challenges, too. They returned to real eating and their other lives the day after Yom Kippur.

But for now, in the crowded Vogel living room, it was thigh against thigh on the peach satin sofa, family flesh expanding or pressed together, a familiar casual touching that went unquestioned by most of them. All this made Elsa Vogel uncomfortable, the closeness of it, the animal-like herding, the physical assumptions. She had always pulled back from all that, flinching from a caress or turning her head aside when someone tried to kiss her, even her husband. Now, her eyes watering slightly, she stood impassively at the doorway and fingered her pearls, stroking them knowingly, one by one, like memories of old lovers, avoiding Julius's restless gaze across the room. Elsa had kept the windows closed tonight because of the rain; a steamy mist had gathered on the inside of the panes; and a terrific concentrated heat came from the radiators, spreading a subtle aroma of family sweat throughout the room. It was becoming more and more unpleasant. It offended Elsa, who did not like the aroma of sweat. Click, she went with her pearls, those rigid, faintly blue, cherished eggs from the seas of Japan that she and Julius had recently decided to sell to help get them through the hard times. Only Elsa could hear the frozen sound of her pearls under her fingers. It was one of the intimate sounds of her life. Click: secretly trying to obliterate everything that was disagreeable or threatening.

While Elsa stood there dreaming for the moment, worrying a little about the glistening chocolate cakes she had baked for her favorite, Oskar, the child-bride Ernestine moved heavily across the room to sit alongside her beloved Lotte. They said a few affectionate words to each other—two unlikely strangers who had been swept into passing family intimacy through marriage—while Lotte took Ernestine's hand in her own. Their fingers twined, an opal ring gleamed, Ernestine smiled gratefully. Like all the Vogels and Manns, Lotte was always kind to Ernestine—her simple, guileless, aging, orphaned girlfriend—and she was always careful to behave in her presence as a reliable and supportive ally. Lotte never criticized and never bullied Ernestine, although she was capable of both, as they all knew; but she was always firm, firmness was strength. That was what Ernestine needed and wanted, it seemed, like a good daughter looking for a good parent. It was Lotte who always brought Ernestine around after the grueling trials of the Yom Kippur fast, after the smelling salts had failed to do the job and Ernestine had to be carried out of the synagogue in a barely conscious state. Lotte held her then, when she felt faint, took her pulse, fed her slowly, spoonful by spoonful, solaced her with sweet whispers, woman to woman; then they parted for another twelve months in an unacknowledged cloud of ambiguous feeling. It was almost romantic, Ernestine sometimes felt, when she had recovered and had a chance to think about it. It made her feel rosy. But it always passed soon enough, too soon, perhaps; at the same time, it was not serious, it hardly counted; and Ernestine, in any case, was dominated by her own prosaically conventional side. There was such a thing, she vaguely understood, as too much worldliness; worldliness could be a threat. It was Selig she cared about, not Lotte. It was Selig who was at the center of her concerns. She worried about his diffidence—there he sat now, in a corner of the room, wearing high-buttoned shoes made of cracked leather, bemused as always, apart from the rest of the family—and about his simple pleasures, stamp-collecting being one of them, the worry persisting in the Vogel living room just under Lotte's attentive, husky monotone and her own girlish smiles; also about Selig's susceptibility to colds and sinus problems, to the croup, to bronchial disturbances in general, even to the threat of tuberculosis. (At the mention of tuberculosis, Ernestine always knocked on wood—her cousin from Leipzig had died of the disease.) Now, whispering into Lotte's ear, enjoying her friend's sharp tangy scent, Ernestine opened her beaded handbag, which she had bought for evenings at the opera, and took out a honey lozenge to pass along to Selig at the other end of the room. Manfred was the messenger.

"I only use one cup of raisins," Lotte, the worldling, was explaining to her friend. "I find one cup more than enough."

"Selig doesn't like raisins," Ernestine said.

"Nonsense," Lotte said. "Everybody likes raisins. It's a question of quality. Where do you buy your raisins?"

There was a pause as Ernestine encouraged Selig from afar with her bright childlike eyes. She nodded at him, he nodded back. Take the lozenge like a good boy, she was saying. Selig understood. They shared life in large part through silent signals, a long-settled physical vocabulary. They had never had much to say aloud to each other; they were not fond of conversation.

"You'll have to come with me the next time I buy raisins," Lotte went on. "I know an extraordinary fruit store, off the beaten track. Benno," she called to her brother-in-law, "where do you buy your raisins?"

"It depends," Benno said, suddenly coming to life on an ottoman across the room. "Boxed or loose? It makes a difference. The raisin people have learned how to keep them fresh in a box. Now, dried apricots are another thing. You have to buy them loose out of the bin if you want . . ." Benno sold ladies' hosiery and similar items wholesale on the road and, even though he was usually home only half the week, was the best cook in the family. They all depended on him for expert advice. His specialties were sauces and desserts à la français, but mainly he cooked only for his friends, who all seemed to be men in love with the domestic arts, including interior design. Benno had his quirks, which the family rarely acknowledged. It made life simpler. Manfred had once come upon him, in his mirrored bedroom, trying on his own merchandise. "Silly, isn't it?" Benno had said, in his tight, adenoidal voice, casually slipping a silk stocking off his hairy leg. "What one doesn't do sometimes," he added, reddening in front of his nephew.

Selig popped another lozenge into his mouth. He loved honey, he loved Ernestine's attentions. "Thank you," he said to Manfred, sucking away in the face of the freezing weather outside and the possibility of the grippe. Uncle Selig and Aunt Ernestine had given Manfred spending money for the trip to Bremerhaven. "For sweets and such things," Ernestine said, as Selig handed over the envelope. "You don't have trouble with sweets, do you? Look at his skin, like a girl's. If I only had your skin. Come here, let me touch you. Don't be afraid. It's only me. Come here."

"Leave him alone," Jenny said, pulling at her corset. "He's not interested in old ladies."

Manfred slipped the money into his pocket. Over his father's head, Kurt winked at him. Jenny, standing alone, wore her sour, dissatisfied look. She had brought Manfred a tie with a polo pony on it, buried in layers of exquisite tissue paper that she immediately took back when Manfred opened the box. "You won't need this where you're going," she said, folding the paper carefully and putting it in her pocketbook. Julius took the tie from his son and held it up to the light, examining it from every angle. The pony was white, the background was brown. It was very sporty. Julius looked sad. The tie depressed him. Everything tonight depressed him. "Pack it up with your other stuff," he said to Manfred. "Be careful. Don't crease it." As for gifts, that was the Vogel side.

On the Mann's, Benno had brought a dozen handkerchiefs, each one monogrammed with Manfred's former initials. Elsa had not had time to warn Benno that Leopold was now somebody else, he was so busy selling underwear and hosiery on the road. LV, the handkerchiefs read, one after the other, in elegant blue thread. "It doesn't matter," Manfred said. "They're beautiful. I love them." Benno blushed. He was always blushing, even at kindness. It was a problem for him when he was selling on the road. He thought it would give him away one day, he was afraid that all those hardheaded, hard-eyed Bavarian retailers in the south, around Munich, who bought his line, would be able to read his mind as he blushed and know once and for all who Benno Mann really was. He had long ago decided that he did not want the matter of his friends and his tastes discussed, he himself would not talk about it. The subject, as perceived by strangers or, worse, those who were close to him, made him cringe. There had been too many jokes, too much malice, over the years in Frankfurt and elsewhere. No one knew that better than Benno Mann. As a result, he was always avoiding confrontations and, sometimes, kindness as well. He had wanted to bring his friend, Willi Eisner, tonight, but Elsa had put her foot down. "It's only family," his sister told him on the phone when he asked about Willi. "That's how I want it. Family only. If we need outsiders we can get our own." Elsa, perhaps, spoke more sharply than she intended, but she said what she meant. Before Willi, who claimed to be a second cousin of Max Reinhardt, there had been Ludwig, Michael, and, years earlier, Sigi. They were all outsiders—Michael, moreover, had been a Swede—and remained outsiders. None of them had ever been accepted by Elsa and the rest. Once again, on the phone, Benno had blushed at his sister's acerbic words and shrugged his family off, smiling cynically to himself.

A sporting tie, handkerchiefs, spending money: it was banal, it was ordinary, it was even dispiriting, but they had to be forgiven, no one was quite sure what Manfred would be allowed to carry out of the country. Everyone needed a tie and handkerchiefs, they reasoned, the authorities would certainly acknowledge that; and the money would be gone by the time he sailed from Bremerhaven. Those choices were probably safe. Finally, Oskar and Lotte—with their usual upstaging flourish—brought out two volumes by Karl May, long after the other gifts had been opened. This turned out to be both banal and perfect, as Kurt muttered to his father out of the side of his mouth. "Leave it to them," he added. "It's Manitou," Manfred dutifully cried at the sight of the dustjackets, referring to the Apache war-chief who was the hero of Karl May's adventure books. "Wonderful," Elsa said to her brother Oskar, with a doting smile. "And I haven't read either of them," Manfred added. Nor would he. He was being disingenuous. He no longer believed in Apache war-chiefs, he was long past Manitou and Karl May.

The gifts were put aside. They made a neat, tiny pile, sitting one on top of the other on the round table in the corner of the room. A few moments passed. Knee touched knee again on the soft peach sofa, flesh swelled in the heat, the families began to circle each other, trying to reposition themselves for another round. Kurt opened a window near Uncle Selig, first making sure there was no draft; the air in the room instantly lightened a bit. Aunt Jenny, who adored her baby brother, Julius, and his children, with only minor reservations, fanned herself with a tiny handkerchief embroidered by her daughter in Munich. Her pince-nez had slipped midway down the bridge of her upturned nose, giving her the appearance of having two sets of eyes. In the dining room, Julius and Elsa kept busy with the wine and cakes, making work for themselves. Julius could not keep still; he was counting the minutes until they would all leave. Benno preened self-consciously on the ottoman he had pulled up and talked about ladies' underwear. The conversation swallowed itself; there was a silence that went on for too long; then Oskar and Lotte decided to raise the subject of American literature. It was a favorite theme with them, as leading family intellectuals. In the sweating room, as they began to hector each other, their words fell on everyone like a distracting summer rain that might wash away for the moment the ordinary debris of daily Frankfurt life. Literature could do that for Oskar and Lotte; all art could, for most of them; they depended on its power.

"There *is* none in America," Oskar began, with his usual sweeping asperity.

"And a shame, too," Lotte cried. "Although of a certain kind of pretense there is no end. The realism of Mr. Lewis sets one's teeth on edge. Such realism comes cheap, it is commonplace, it exists everywhere, on any street corner for all to see. And they speak of him as a Nobel possibility. But the American imagination, where is that? Where is the noble spirit, the imagery, the bright bird of fancy we all seek?" Her eyes opened wide in her knobby face, perhaps in admiration of her own words. "There is something hard in that country, I've always known it, something stonelike that forces poetry aside."

"I don't want to talk about America tonight," Manfred said.

"We know so little American writing here," Julius added stiffly, checking his watch.

"Because there *is* so little," Oskar said.

"Come now," Julius admonished. "Don't exaggerate."

"I read *The Good Earth*," Elsa said. "It seemed to me very fine."

"Bourgeois cluckings about starving chinks," Oskar stated.

"What did he say?" Ernestine asked.

Elsa frowned. "I loved *The Good Earth*."

"If the truth were known," Lotte said, "so did I."

"A hundred and fifty years from now in America, what will be read?" Oskar asked.

"Ernest Hemingway," Kurt said.

"Child's play."

"Do we have to talk about America tonight?" Manfred asked. Ah-mer-i-kha, he pronounced it, with a deliberately ragged, perverse edge.

"I didn't read *The Good Earth*," Selig finally brought out, wheezing in his corner. "But I think Ernest Hemingway should be writing Indian stories, like Karl May." After coughing once at this challenge, he subsided in his chair. Across the room, his wife looked as if she was considering his remark, then closed her eyes. Ernestine was pleasantly bored. Lotte and Oskar knew how to talk about everything. Now she would try to nap for half a minute through all the words.

"What about Mark Twain?" Manfred asked.

"That's right, darling," Aunt Lotte said. "Everybody loves Mark Twain."

"And what about Hawthorne and Melville?" Kurt asked.

"I tried to read that book about the whale once," Benno said, shuddering.

"And Fenimore Cooper?" Kurt went on, beginning to warm to the subject. "Fenimore Cooper makes Karl May look like Felix Salten with his dear little anthropomorphic fawns and their sensitive German offspring. That's what I call provincialism. German provincialism and German sentimentality. The ghastly Teutonic twins. What a combination."

Oskar was not used to being contradicted, especially by his family. He swallowed hard and looked stern. It was his Bourse look. "We've all read Mark Twain," he said to his nephew. "Even Goebbels has read Mark Twain. But naturally, you've missed my point. I'm not talking about writing, per se. About mere writers, about words on a page. I'm talking about civilization overall, about literature, art, music, drama, a nation's culture in its totality, that 'thing' that becomes the essence of its soul, and, if I may speak from my own authority—if, of course, you have no objection—about a whole overrated industrial structure that is now coming apart at a rate that would shock you. I know what I say. I know it from the inside. America will never come out of this depression. The unemployment numbers far exceed our own. The GNP is shockingly down. People starve there. There will be no recovery. That is now a European assumption. All the money markets know it. Paris, Brussels, Geneva. Moscow counts on it."

"Please," Elsa begged, acknowledging Manfred with a nod of her head.

"You're changing the subject," Kurt said to his uncle. "You always change the subject when you don't like the way the conversation is going."

"You have to admit, Oskar, darling . . ." Lotte began archly.

"I don't have to admit anything," Oskar said, dismissing them all with a suddenly bored look. He sat back then and reached for one of Elsa's cakes, spilling crumbs onto his lap as he bit into it. Oskar had huge, swelling thighs, a definitive Mann trait, carried and transmitted by both his Mann grandparents and, in turn, by theirs. It had been passed on to him intact. Elsa shared it, although modestly, while Benno, who prided himself on his daily exercises, had worked obsessively at slimming down the natural Mann bulk to a sleek, hard shape. As a child, Manfred had straddled Uncle Oskar's formidable thighs, playing silly games, until his uncle, whose mind always seemed to be on something else, grew tired of him. Then he had begun to use Benno's legs, slicing his buttocks sometimes on the very edge of Benno's sharp thighbones, bouncing in a passion of sensual pleasure, as he pretended to be a cowboy and Uncle Benno his horse. (By that time, Benno, who did not believe in half-measures, had discovered that he was beginning to go bald and shaved all his hair off; his head had remained clean-shaven ever since.) The games had stopped the day Manfred noticed a transfixed and, at the same time, alarmed look cross his uncle's face as he sat on his lap, facing him, a powerful, trancelike expression that vanished in an instant but left his uncle—and himself—shaken.

"In any case," Benno now remarked to his brother, "no one reads anymore. No one in Germany, at least." A wire-thin bracelet on his left wrist, a gift from Willi, glinted in the pink light. "Everybody goes to the movies. It's the only safe place. It's dark in there, nobody knows who you are, nobody even knows you're there."

"It destroys the soul," Lotte said. Her intelligent face, with its short straight hair framing the sharp bones, challenged them all.

"What does?" Benno asked.

"The movies. American movies especially. Images drive out words. Thinking evaporates. The mind feeds on pictures, forgets ideas. That's how early man lived, that's how prehistory worked, that's primitivism."

"I haven't been to a movie in six months," Julius said.

"Movies will be the death of language," Lotte continued. "Literacy will disappear. Words will be erased from the consciousness. Books will go first." She stroked Ernestine's hand firmly, waking her.

"That's a standard argument," Kurt said. "Everybody's heard it before."

"I beg you, no politics," Elsa said, looking to Oskar for help.

"American movies kill the soul, French movies corrupt the spirit. I stand by my opinion."

"And German films?" Kurt asked.

"Jew Suss, Ohm Kruger, Olga Tschechowa, and Zara Leander with the big . . ." and Manfred shaped a full-bodied figure in the air.

"Listen to him," Jenny said.

"All the good ones are in America," Manfred explained. "They all went to Hollywood."

"Hollywood," Jenny repeated, dreamily.

"I beg of you," Elsa insisted, pointing to the walls.

"*Die Alte* again," Manfred said.

"Who is *die Alte?*" Ernestine asked.

"*Die Alte* is *die Alte*," Selig told his wife. "She's *die Alte* next door."

No one was ever quite sure how to take Selig, he was sometimes so obvious. All his life, saying things like that, which somehow always took everyone by surprise, he had never been sure himself of how serious he was, asserting himself forcefully one moment, pulling back timidly the next, vague and contradictory, as though there might be an embarrassing penalty for pushing one's self ahead, for "arrogance," as he saw it. He was something like his brother, Julius, in that; neither liked to impose or make his presence felt arbitrarily, neither could take his own words—or others'—without a degree of habitual irony.

"Are you talking about Frau Schuster?" Lotte asked. "The dwarf?"

"Please," Elsa said. "Words have wings."

"I just asked if *die Alte* is Frau Schuster."

"You called her a dwarf," Oskar said. "It's indiscreet. She lives on the other side of the wall. She might hear. All of us have to be more discreet. We have to try twice as hard as anybody else."

"Exactly," Elsa affirmed.

"The only way we could be more discreet," Kurt said, "is if we all stopped breathing."

"You know what I mean," Oskar said.

"What *do* you mean?" Jenny asked. "My pension has been cut in half. It's the same for all Jewish widows. The War Office says it will be taken away altogether in six months. How should I be more discreet?"

"This conversation will get us nowhere," Benno said. He was rocking back and forth on the ottoman with his hands clasped around his knees.

"Somebody explain to me," Jenny said.

"I don't want to talk about it," Benno said. "It's grim and I don't like to feel grim, it makes me look ugly. I live with it all day long, like everybody else, and I'm tired of it." Nobody answered him. "What would make me happy," he continued, "is to see my brother, Oskar, smile. Once. What does it cost? A simple smile. It wouldn't hurt you either, Selig, with all due respect. That's right, just a bit more at the corners . . . There, it's almost a smile. Anything for a celebration is what I say. Even a catastrophe will do. At any excuse, Willi opens the champagne and we drink up. My nephew is going to America with a new name. Hail Christoforo Columbo, Amerigo Vespucci, hail Manfredo Vogello!" And he lifted his glass of wine with the rest of them, while they all drank. Now it was Manfred's turn to blush, and as he felt the blood rise, he lowered his eyes. He was wishing the obvious, that they would forget about him tonight. "That's more like it," his uncle cried. "We need all the fresh pink cheeks we can get in this family."

"Manfredo Vogello, indeed," Lotte said, turning to her nephew. "I'll take you under any name."

"Why don't you come to America?" Manfred asked her, impulsively.

"Ah," she said, longingly. "With my hypertension." She laughed again. "Don't mind me," she went on. "I was made for Frankfurt, Frankfurt was made for me, and here I stay, good health, bad health. Besides, I'm German, whatever else people have the bad taste to insist, and I'm also getting to be an old woman, like everybody around me. It's too late."

"How old are you, actually?" Jenny asked, examining her fingernails.

"Wouldn't you like to know," Lotte said.

"How old is she?" Jenny asked Elsa, smiling at last.

"Do you really think," Ernestine began, "that movies destroy the soul?"

"Come on, confess," Jenny said. "How old?"

"Ginger Rogers, too?" Ernestine asked.

"I prefer Carole Lombard," Manfred said, solemnly.

Ernestine, after a comfortable silence, began to offer Lotte another recipe. She counted out the weights and measures on her fingers. "No sugar," she said. "Just noodles." Recipes littered her head, as they littered her sister-in-law's, like arithmetic problems left over from childhood. She could put them together accurately at a moment's notice. They all could. "But I never use sugar," Lotte answered. "Since last year's report in the *Zeitung* . . ."

In a corner, Jenny and Selig were now deep in family conversation. They sat side by side on two straight-backed chairs, without looking at each other. They had sat on twin commodes talking to each other like that when they were children. Jenny had found a new treatment for hemorrhoids, she announced, facing the rest of the room. She spoke out of the side of her mouth so only Selig could hear. The treatment eased the pain, cut the swelling, stopped the bleeding, and was not messy, awkward, or personally embarrassing. "You know what I mean," she said to her brother. It all had to do with a special softening process, she added. "You understand?" she asked. "You take it orally." As she offered up each detail, she played with her pince-nez for emphasis, snapping it on and off the soft bridge of her nose. "It works," she told Selig. "You can take my word for it."

Selig listened anxiously to his sister's words, gazing in front of him. They seemed to offer some hope. His hemorrhoids were like fat grapes. He had also begun to awaken occasionally at night with a numb left arm. Sometimes he developed a leg cramp after walking only a block, and for that he had stopped smoking, at Dr. Schwabacher's insistence. He yearned for a cigarette now, a Gitane or a Gauloise, something French with a delicious barnyard scent to it. To hold between his fingers and hide behind the beautiful smoke as it floated lazily to the ceiling, then to inhale greedily, like a guilty boy. But Selig couldn't afford such fantasies. They would help to kill him, Dr. S. had promised. With an effort, he turned his attention to Jenny again. She was talking about martyrdom, her eyes popping. Again she snapped her pince-nez on and stared straight ahead. There were not many war widows with memories as staunch as Jenny's. She had been celibate for almost twenty years, ever since the Somme. Selig knew the story well. It was tragic even though it was commonplace; all he had to do was think of *die Alte* next door; she, too, had lost her husband late in the war. Selig had been listening to his sister for almost half a century, on countless subjects of her own choosing, and she was still so compelling that she could make him forget his bronchitis, his hemorrhoids, and his need for a Gauloise. She talked on, describing again a recent visit to her daughter in Munich. Lisl was developing a phobia, she said with a slight groan, about leaving her house at night; she was afraid of being abducted by the Nazis and carried away to one of the nearby concentration camps; apparently it had happened to one of her friends. Ah, Selig said, tucking his feet under the chair. Like Jenny, he gave a slight groan. It was a strange time for everyone, he said in a helpless voice, poor girl, no one was safe, no one was immune, not one of them; and sitting there silently, his temples throbbing, he began to feel guilty about his niece.

There was a flurry then on the other side of the room as Kurt tried to engage Oskar again in the argument about American Kultur, but Oskar would have no part of it. He had already had his say. Kurt could take it or leave it. There were other things to discuss. Who, for example, did Kurt think the Prince of Wales should marry? His choice could make a difference to the world. Those things still counted for something in the money markets and elsewhere, even in America, even in the Soviet. Oh, the Prince of Wales, Kurt said, waving the subject away. It's too silly, it's not for adults. No, Oskar insisted, grinning now as he put his gruffness aside, it could change the course of Empire. It really could. The whole world would be affected. Oh, come, Kurt said, it's for shopgirls, not for us. So you don't think it's important, his uncle asked. No, I don't, Kurt said. The Prince of Wales is for Tante Ernestine. They began to laugh together and look for something else to argue about. They had been at it since Kurt entered Gymnasium, years before. They were friends; in the end, they always enjoyed each other. Elsa Vogel listened to them laugh. The sound pleased her, it made her feel calm. Everything was still intact. She had tried hard to confine the evening within domestic limits, and she had succeeded. Oskar had a wonderful laugh, so did Kurt. They even sounded alike, her brother, her son. She began to laugh with them, but only to herself, and from nervous habit set about serving a second round of cakes, which she eyed warily in the overheated room; the chocolate had begun to run. Behind her, Manfred—the subject everyone had been more or less avoiding all evening—continued to stroll from one part of the room to another, sitting in uninvited, biting his nails, eavesdropping, forgetting everything he heard a moment later. "What's so funny?" he asked his mother, without waiting for an answer. He moved off again, away from her smile and the sound of his brother and uncle laughing together. He was neatly cocooned inside himself, cushioned in perfect isolation from everyone else. Nothing could happen to him through all the talk. He had made sure of that tonight. His father was passing out additional silverware, just to keep busy. The gap in his right hand made a great U-shape. With his father,

Manfred shared the gift of protective coloration. They were masters of the same techniques, father and son, outside it all when they wanted to be, barely visible, untouchable for the evening.

"And how is business?" Oskar called across the room to Selig. He was serious again.

"There is no business." This was an exaggeration, but true enough. They both understood that it was a way of talking about the new reality. Selig, being a lawyer, could only have Jewish clients, and his Jewish clients, being Jewish, were losing their businesses. Therefore . . .

"It will calm down soon."

"You think so?" Selig asked. He sounded skeptical. "Seems to me any calmer and we'll be stranded on the shoals." He laughed without making a sound. "I mean the ship of affairs."

"There will be changes," Oskar said to him. "Things will be back to normal. The ship of affairs will take sail. There's an international momentum in these things that can't be resisted. The money markets won't allow. . ."

Ernestine fluttered her hands at their words. She couldn't bear the conversation. It was beyond her, yet, at the same time, remained perfectly clear. Bits of old recipes came to her mind unbidden, weights and measures, forgotten ingredients. "How are we going to live if there's no money?" she asked.

Lotte reached for her hand and stroked it, exactly as though it were a damaged bird.

"The world goes on living," Benno said, in his dry voice. He threw his head back dramatically, his bare skull shining pink. "The world's like a concertina," he improvised, staring upwards, as though he were looking for the words on the ceiling. "It contracts, it expands, contracts, expands, always breathing in and out in the struggle to stay alive. Like all organisms, big or small. Now it contracts. It folds in on itself. For the moment, there is no breathing space. It's a phase, part of a terrible cycle. The wrong time to be alive. You don't think so, Kurt?" His nephew was eyeing him dubiously. "Think about it," he suggested, as Kurt remained silent. "Good, bad, in, out, expand, contract, it's the universal rhythm, it's the same everywhere."

"What do we know about everywhere, here in Frankfurt?" Kurt asked.

"I can't seem to get it in my head about Leo's new name," Ernestine said.

"Manfred," her nephew stated.

"That's what I mean," Ernestine said, raising a laugh again.

"Does anyone know why Frau Göring lost her belief in church?" Manfred asked in desperation.

"Because," Oskar said, "she lost her faith in the resurrection of the flesh. And you got it a little mixed-up. You told it wrong."

"Don't go teaching him how to tell jokes," Kurt said. "He doesn't need any encouragement."

"We went to services in the West End shul this morning," Ernestine said.

"I don't like the rabbi there," Elsa said. "I never did."

"I prefer him to what's-his-name at the Borneplatz."

"You should try the Friedberger Anlage," Selig suggested. "I've been telling you that for years."

"I saw the Rothschilds there on Passover."

"Impossible."

"It was one of the cousins. Mayr and Bettina."

"Impossible," Oskar repeated. "They weren't even in Frankfurt for Passover. I know that for a fact."

"I hate shul," Manfred said.

"The West End shul looks like Baghdad to me," Julius stated. "Every time I go there I feel that I'm in the Orient."

"Well, the Friedberger looks like a church," Benno added.

"I didn't know you cared," Oskar said.

"I don't, I'm making an architectural comment."

"Being an agnostic simplifies life," Lotte said. "You never have to make those choices." Ernestine yawned, then folded her tiny hands in her lap. "I read," Julius said, "that Yehudi Menuhin is going to retire." "But he's still only a baby." "He says he's sick of playing concerts. He says he's sick of the world. He says he's going to hide behind a Chinese wall three miles long." "He's going to China?" Ernestine asked. Gieseking would be in Frankfurt next week, someone reminded them. Debussy and a little Beethoven. Which Beethoven, Lotte asked. It made a difference to her. If it were early Beethoven, she'd spend Oskar's money elsewhere. There was going to be a new *Tosca* at the opera house in February. Manfred would miss it. A shame: he sometimes went to the opera with Oskar and Lotte. He liked the opera. He liked *Tosca*. There were rumors of Cebotari or Giannini for the cast, maybe Gigli, too. Now Manfred yawned. It shut out all the sound around him, made the inside of his head resonate like a seashell. After a moment, he reached into his pocket to make sure that Adele Gordon's letter was still there. Adele Gordon said that he would have to decide between boy-girl or all-boy. Your Highness, she said. The lady of the house likes things her way. Evidently some people were afraid of girls. Paradoxes and mysteries. He liked paradoxes, too. Through the humidity, the doorbell finally rang.

It was Dr. Schwabacher, come to say good-bye. A frown crossed Benno's face when he saw the doctor walk in. Dr. Schwabacher was an outsider, like Willi. His neat little goatee was damp from the drizzle outside. Tiny drops of water hung there, glistening. The family stood up as he came into the room. He was in a hurry, he told them, urging them to sit down in his professional voice. Ernestine made way for him. "Don't you go catching cold," she said. "That would never do." "There you are," Dr. S. called to Manfred, reaching out to hold him by the shoulders. By now they were the same height. "Look me in the eye," Dr. S. said. "No, no, no," he said to Elsa, who stood at his right elbow, plate in hand, "nothing for me, thank you. I was just passing by. I had to see my friend here. Now look me in the eye," he repeated. Manfred looked him in the eye, fastened to the old doctor by childhood discipline. "Excited? Nervous? Anxious?" Dr. S. looked very serious as he asked the questions. Yes, Manfred admitted, under the doctor's stern gaze, he was all those things. "It's only natural," Dr. S. said, "it's a sign of life. A sign of human consciousness. I brought you something," he then added, maneuvering the patient into a corner by the shoulders. "Here." He brought out a tiny vial made of blue glass. Manfred stared at it before taking it from the doctor. "When your stomach bothers you, if you're nervous or too excited, you know what I mean, even for a headache, with your history, you put a drop or two on your fingertip and swallow it. It will do the trick every time. It's good for at least six months." "What is it?" Manfred asked. "It's an opium compound," Dr. S. said. "Considerably diluted. Don't worry. It's safe. There's nothing to be afraid of. Use it when you need it. I always keep a bottle at home for myself," he went on, "but in a more concentrated form. I save it for my own variety of stomach ailment." Dr. S. laughed. "Sometimes I create one just for the pleasure of curing it. It'll help you get through the first six months, until you're settled in, until you feel at home with Uncle Sam." Dr. S. laughed again. "You know there's more where that came from."

They were all on their feet. Julius smiled at the movement around him.

"Anything new with Elsa?" Dr. Schwabacher asked him, after Manfred had pocketed the little blue bottle. Julius shook his head. There had been nothing new with Elsa for months. "Patience," Dr. S. suggested, in a muted voice. Ernestine approached the doctor then, her head tilted to one side, flirting a little. She wanted to talk to him about Selig's sugar count. Lotte, just behind her, had blood pressure on her mind. Only Oskar, among them, avoided Dr. S. Dr. Schwabacher was not his doctor, and, in any case, he had no use for the medical profession. He was trying now to get Manfred into a corner himself. "You made me very unhappy tonight," Benno was saying to Elsa, standing in the middle of the room. "What are you talking about?" she asked. "No outsiders, you said." "So?" Elsa answered. "Then what is Hans Schwabacher doing here? He's as much an outsider as Willi." "Stop it, Benno," she said fiercely. "This party wasn't for you tonight." "Still," he said. He began to blush. "Stop it, I say," his sister hissed at him. "On such an occasion . . ."

"Manfred," Oskar called, "do you have an address book? Come over here and bring your address book." "Do as Uncle Oskar says," Elsa told her son. "Hurry." The English king, Dr. S. informed the room in grave tones, was seriously ill. He had heard it on the radio tonight. "I told you," Oskar said to Kurt, over his shoulder. There were special bulletins every three hours from the palace, Dr. S. said. It was clear that he couldn't possibly last out the month. The special bulletins camouflaged the symptoms, but one could read between the lines . . . Was that good or bad, Lotte wanted to know. What would it mean to Germany if the English king died? Kurt raised his eyebrows at the news, considering the question in terms of his visa application. "Give me that address book," Oskar ordered. He pushed Manfred into a corner, slipped a twenty-mark note into his pocket. "No," Manfred protested. "Don't be a fool," Oskar said. "Now listen to me, because I only say things once. You and Kurt are going to be my heirs. You hear? After Tante Lotte, of course. You and Kurt equal. Do you understand what I'm talking about? Stop grinding your teeth like that. Where'd you get such a habit? Are you a Communist? Tell me the truth. Because if you're a Communist. Now listen. Part of my money is in Switzerland. In Zurich. In a Zurich bank. Give me that book so I can write the address in and the name of my manager. It'll be under *S*. Remember that and don't lose it. Under *S*. for Switzerland. Who knows when it'll come in handy? Do you understand what I'm saying to you? I wonder. *Are* you a Communist? You can't be a Communist, I would have known, I can always tell. Here, I'll write it down for you. I need a pencil. Get me a pencil."

Benno came over while Oskar was scribbling in Manfred's address book and slipped ten marks into Manfred's pocket. "No," Manfred protested. "Let me give you a kiss," Benno said. "One little kiss, for God's sake, don't be so maidenly." Somebody else was also kissing him. "What did Uncle Oskar want?" Elsa whispered into her son's ear. She was smiling. "You can tell me," she said. On the ottoman, Selig was putting on his rubbers. There were sweat stains under his arms, perfect black circles that appeared every time he bent over to tug at his shoes. Oh, dear, Ernestine sighed, turning around aimlessly in the middle of the room, holding a bag of Elsa's cakes in her hand to take home. Jenny pulled Manfred to her, her arms caught him. He was trapped again. Enfolded like that, he could smell her talc, her cologne, her dim crotch smell. No one else smelled like Jenny, who imported her scents from Paris. He clung to her, catching a glimpse of her powerful Wilhelmian cleavage, which was celebrated in the family. Even Benno admired it. She had wide, spreading nipples, Manfred knew, two inches across, which looked like purple plums splattered in the light of her bedroom. He had found her there more than once, naked from the waist up, placid and expectant as she dressed in front of the vanity table. She enveloped him now, breathing in his own smell. "Oh," she said, in a stubborn voice, refusing to let go. "Well, darling," Lotte said, sounding hoarse. "Whose umbrella is this?" Julius asked, holding one up. He was already out on the landing, ready to see them off. "All the umbrellas are still wet." "You won't forget this little bottle now?" Dr. S. said. "You can always get a refill. I'm always here to help." "You've got it now," Oskar said, thrusting Manfred's address book into his hand. "Hold onto it. Maybe it'll save your life some day. You never know." "Well, if that's what you call a kiss," Benno complained, with fake indignation. "You should really wear a hat in this weather," Kurt said to his uncle. "It's ridiculous for you to go out like that." "You're right," Benno said, touching his bare skull. "I keep forgetting." "What do you mean, you keep forgetting?" Oskar said. "You don't even own a hat." They were moving through the door. There was a lot of noise on the landing. Julius stood alongside Manfred, his arm around his son's shoulder. They looked alike, thin, fine noses, small feet. Elsa was kissing Benno. Kurt had the widow Jenny in his arms. Had anyone heard the shooting tonight from the other side of the river? Dr. S. asked. "Oh, shooting," Ernestine said, her hand at her throat. It was the second time in a week, the doctor added. Lotte gave a shudder. "Come on," she said to Oskar. "It's still raining outside." Ernestine was suddenly weeping, out of nowhere. "Oh my God, none of that," Lotte said. "You promised. Watch your step now. The stairs are slippery." "Come," Selig said, taking Ernestine's arm. "Control yourself." Elsa clutched at her pearls. "Shh," she said. "It's late." She gave them a disapproving look. Julius gripped Manfred's shoulder. Good-bye, someone called, starting down the stairs. "Be careful," Lotte said. "Are you all right?" Selig asked Ernestine. "Are you sure?" Then they were all going down the stairs. Someone hiccuped. Benno blew a kiss at Manfred. "Give my regards to Broadway," he called.

Good-bye. So long. Wiedersehen, mein liebling.

Slam, went the door of history in their faces.

Chapter Five

From the ship's railing, where Manfred stood shivering, Kurt's face grew smaller and smaller, shrinking into itself as the crowd swallowed it up among all the other faces. Soon he looked like everybody else. Manfred could just make out the broad planes of his perfectly white cheekbones, the stark black eyebrows, arched in perpetual doubt, the triangular Mann chin, so much like Oskar's. An unfamiliar, anguished look nested in his squinting eyes today. Yet Kurt was cheering, like all the rest. He wore a black overcoat, with a velvet collar, and a white silk scarf around his neck, like an aviator. Manfred's own coat was grey herringbone wool, belted and matched to his expensive knickers, which bore hardly a trace of the misadventure near the Schillerplatz. Manfred also had on a new beret. He could feel a steady pull around him as he stood at the rail, trying to keep Kurt in focus, as though an invisible tide was at work, a constant tremor and hum that seemed to come from right under the soles of his feet. He realized that the ship was moving. It had been moving for five minutes. Frothy water churned beneath him, the ship's gangplank lay on its side on the dock, near Kurt.

They had had twenty-four hours together, shared on the icy train trip to North Bremen, as the agency had suggested, then overnight in a good commercial hotel in the same vast double bed. Manfred had never before shared a bed with his brother. The idea had made them both timid. They had crept in self-consciously, avoiding the other's eyes, and had lain on the outer edges of the mattress, on either side, afraid they might touch each other. The sound of a clarinet playing imitation jazz came from a nightclub across the street. Kurt kept humming along with it, off-key, until he fell asleep. Once, during the night, Manfred's foot somehow found Kurt's side of the cold sheet, twitching Manfred awake. Earlier in the evening, they had eaten an enormous dinner in a dark smoky basement restaurant, recommended by Uncle Oskar, that was filled with hearty maritime types with ruddy complexions and loud seagoing voices. Dinner was a success, they agreed; you couldn't find a restaurant like that in Frankfurt-am-Main; Oskar was always reliable about such matters. They had seafood from the Baltic, steaks from Holstein, and two bottles of Rhone wine. There had even been a couple of whores at the bar, eyeing the brothers. Manfred had been drunk. He had talked a lot, through what seemed to him to be a hole in the top of his skull. So had Kurt, their voices cutting over each other through the clatter of dishes and the sound of rowdy singing around them. Kurt had a brandy and a fat Cuban cigar after dinner, Manfred ate a second cream pastry. They talked until midnight, before going back to the hotel; at the very end of the evening, they toasted a long life to Aunt Lotte. They were laughing at one of Manfred's terrible jokes as they rode up in the elevator to their stately room.

Before turning out the light, Kurt brought out a brown parcel and unwrapped it for Manfred. It was a *Cassell's German-English Dictionary*, second edition. "Ah," Manfred said with pleasure, hefting the volume in one hand. "It's Breul's," Kurt explained. He had to sit down first to keep his balance. "Not Miss Weir's, whoever she was. Did you know that Breul was a Doctor of Literature from Cambridge? And Reader in Germanic? That makes the difference. That's the horse's mouth, being at Cambridge. His Berlin degree means nothing. German academicism, that's all. I considered the new Grieg-Schroer and even the Glugel-Schmidt-Tanger. Neither was worthy of the occasion, you might say. I also thought maybe the small Muret would be more convenient, because of its size, but, in the end, after trying the Heyne, the Kluge, *and* the Saalfeld, which itself is not so bad, one of the best, in fact, I decided that the Cassell was far and away the most practical, the most complete, the most useful, and . . ." By then, Manfred's eyes were beginning to close. He was sitting on his bed in his underwear, facing his brother. The dictionary lay in his lap. "You have your visa, your passport, all the instructions?" Kurt snapped at him. Manfred pointed to a canvas shoulder bag hanging on the back of the desk chair. "Don't lose anything," Kurt ordered. "I want to be sure to see the last of you tomorrow."

"Oh, shut up," Manfred said sleepily.

In the morning, they arrived at the dock early. They were there an hour before they had to be. All the Vogels arrived early, wherever they went; it was a family joke. The Scranton Boy was also there at the dock, sitting on his luggage while he waited to board ship. His parents stood behind him, fidgeting. The Scranton Boy kept falling asleep right where he sat, chin flopping onto his chest, his nose running in the cold. They were soon joined by the two Cleveland Boys. The Cleveland Boys had red hair, with thick freckles sloping down identically on each side of their noses. Their skin was pink, so were their eyebrows. It also looked to Manfred as if they had pinkeye, like a rabbit he had kept with Joachim Adler in a corner of the school lab the semester before. They were brothers, one of them announced. Their name was Heller. "Samuel," he added. "And Louis," pointing to the shorter one. "OK, Sam," Kurt said. "Samuel," the redhead answered. Everyone shook hands limply, at Kurt's instigation, Scranton Boy, the two Cleveland Boys, and Manfred. For a while, they all thought that the Louisville Boy had reneged, that he had lost his nerve or his way or his visa, but he finally showed up a half-hour late, out of breath, alone, disheveled, pieces of his clothes hanging out of the corners of his cardboard suitcase. "Jesus Christ," he moaned at the commotion around him. He wore a card pinned to the lapel of his coat that read "My name is Hugo Bondi."

On the dock, officials shouted at each other, porters ran around with their metal handcarts, looking for business, a crane was lifting cargo into the hold. Somewhere in the middle distance, Manfred could see a massive, upward-curving section of shining black hull, spotted by portholes. Only part of his mind recognized it as the *Europa*. Kurt bought them all coffee from a vendor. "This will help to keep you alive," he said, offering the hot paper cups around. Then the Scranton Boy's father bought them sweet buns to go with the coffee. At the very edge of their tight little cluster, a social worker from the Jewish Welfare Agency stood around with a clipboard in his hand. He told them to line up alphabetically, count to five, then fall out. Slowly, his breath steaming, he finished his coffee and a second sweet bun; his eyes were swollen from lack of sleep. A few minutes later, he told them to line up again. Up ahead of him, Manfred could see two heads of red hair, one an inch taller than the other. Hugo Bondi, taller than all of them, headed the line. *Eins*, he shouted above the noise. Then the others followed along, *zwei, drei, vier, fünf*. Manfred was last. Alongside them, while the official business went on, Kurt chatted with the Scranton Boy's parents. They were tentative with each other, trying to make neutral conversation. Again, Manfred caught a glimpse of the ship's hull broadside; it hardly registered. They checked and double-checked their identity papers, their itinerary, the names of their new

American families. At each response, the social worker made a careful mark on his clipboard. Friedman, Hugo Bondi called out. Wise, the Heller brothers said. Then Schoenberg in Scranton and Gordon in Baltimore. Kurt had a brief, sharp exchange with the social worker about the unhealthy effect on the boys of standing around in the cold and was nervously backed by the Scranton Boy's parents. "There must be another way," Scranton Boy's father said over and over in a grinding monotone that finally ended up making Kurt even more irritable. "I am not a miracle worker," the agency representative said, in a distant voice. It was a statement that he was forced to make every day of his life. "These things take their prescribed course. I am doing my best. I assure you there are no shortcuts. It might have helped if you had shown up at the appropriate time. I can't be responsible for everything." Manfred listened to them squabble and scratched his arm. Like Kurt, he was hung over. The soured taste of Rhone wine was still in his mouth. He thought he could smell Kurt's brandy breath. Jesus Christ, Hugo Bondi kept saying, swallowing air. His chest heaved, he gasped. Witch's tit, he muttered under the other sounds. "What'd you say?" Kurt asked nervously. "Speak up." "Nothing," Hugo Bondi said. Kurt gave him a sly look. "Any relation to Somerset Maugham, by any chance?" "Who?" "Of Hugo Bondige," Kurt said in English, with a horrible grin. "Please," Manfred groaned. His teeth were chattering. In a corner, the Cleveland brothers stuck together silently. They spoke only to each other, in a formal way, while the Scranton Boy, trying to avoid his mother and father, stayed at Manfred's side. His father was still complaining. "It's a damn shame," he kept repeating. "I'll report it. It's impossible. In this weather." Nobody was paying attention. His steamy breath puffed out in angry little clouds. "I think I'm going to faint," Scranton Boy whispered to Manfred. Their upper arms touched. "Don't move," Scranton Boy said. "Stay with me."

At last, one by one, they began to climb the gangplank. It was angled much more sharply than Manfred had ever imagined in his landlocked dreams about the *Europa*. His canvas bag banged against his hip. Gravity pulled at him. Climbing hard, he eyed the portholes, the gleaming hull, the filthy water below. A rainbow slick floated between the ship and the dock. Up front, Hugo Bondi and the social worker led the way. Then came Samuel and Louis Heller, followed by the Scranton Boy and Manfred. Kurt and the Scranton Boy's parents were somewhere behind them, holding on to their visitor's passes. Once on board, the boys regrouped, counting out loud again for the social worker's benefit. *Eins*, shouted Hugo Bondi. Then the others shouted their own numbers after him, *zwei, drei, vier, funf*. Still in single file, they moved along the deck, as ordered. Stay together, the social worker called out. Manfred stepped on the Scranton Boy's heels, they both stumbled. This way, the social worker called. They made a right turn and went sideways down a steep flight of metal steps, which eventually led them, after another turn, through a narrow corridor to their own tiny cabin. Count off, the social worker cried, hurry, while Manfred again stumbled against the Scranton Boy. *Funf*, he called, feeling for his canvas bag. They shuffled into the cabin hesitantly, then, coming to rest together for a moment in the middle of the floor, staring around them strangely, sniffing, blinking, appraising the bunks, the bare overhead light, caged in wire, the worn linoleum beneath their feet, the single washbasin, the muffled closeness that hovered around them. They were deep inside the skeleton of the ship, dug in like bone marrow. Scranton Boy held onto Manfred's sleeve. "Where's the window?" Hugo Bondi asked, looking at the blank walls. At the social worker's orders, they stashed their belongings in upright lockers and had their papers stamped again. There were doubledeckers in the cabin; one bunk would be empty. Kurt and Scranton Boy's parents stood at the door of the cabin, watching them settle in. "If you please," Scranton Boy's mother called out, "perhaps you would be so kind as to let him have a lower bed. Yes? He suffers from vertigo." At her words, Scranton Boy turned red and looked away. The morning's hot porridge kept rising in Manfred's throat, through the taste of the Rhone wine. Even the sweet bun hadn't wiped that out. Outside, in the corridor, a gong rang twice. "All right boys," the social worker said. "Ten minutes."

Up on deck again, Kurt pulled Manfred to him in a harsh, unexpected tug that took Manfred's breath away. His white silk scarf was cold against Manfred's cheek. "Write to me," Kurt said into Manfred's ear. "In English. Forget German. It's worthless. You have a new dictionary. Use it. You're a smart boy." "Don't go yet," Manfred said, as he felt Kurt begin to pull away from him. He pulled back at his brother. "Please," Manfred said. "You know how to spell my name," Kurt said, tilting Manfred's beret at a rakish angle. "You know where I live." He traced the bridge of Manfred's nose with his forefinger. "Wait," Manfred said. A bell rang. There was a scurrying for the gangplank. "All right, children," the social worker said, making one last check mark on his clipboard. "Children, my ass," Hugo Bondi mumbled under his breath. Scranton Boy's father suddenly lowered his head and covered his eyes with his hand, just as Julius Vogel had done the day before at the Frankfurt train station. Stroking his shoulder, his wife made strange sounds. Manfred stared at them. They began to careen around the deck together, locked in each other's arms, as though they couldn't control their muscles. "Oh, my God," Scranton Boy said, in a small voice, watching them. Overhead now, the winter sun hung bleached in the morning haze. The wind grew stronger coming from downriver. The ship was huge, Manfred could not take it in. There was another blast, this time from a funnel somewhere overhead. "I hope it doesn't rain," he heard himself say. "Remember," Kurt breathed into his ear. "Wait," Manfred said, reaching out. Then Kurt suddenly disappeared, the parents were gone, the social worker was gone with them. "Here," Hugo Bondi called, beckoning to them from the railing. "Over here."

They grabbed the space, from which they could see the dock below. Everyone was waving white handkerchiefs. A line of idle porters waited behind the crowd, standing alongside their empty handcarts. Some confetti fell limply from an upper deck. Near the boys, a couple of trim sailors in immaculate uniforms worked easily, coiling ropes. A few officers stood by, their arms folded, making jokes to each other. "If I faint," Scranton Boy said, "you'll find smelling salts in the pocket of my coat." He kept touching Manfred. "You won't faint," Manfred said, trying to find Kurt in the crowd on the dock below. His eyes worked from left to right, swiftly, back to front, one face at a time. He saw him at last, jammed in against the barrier. His white scarf looked as if it was about to fall off. Kurt was arguing with a woman who stood alongside him. Manfred could see the woman clearly, he could see everything clearly. She wore a tight felt cloche with a green feather in it. Kurt shouted something at her, making an angry gesture. The woman shouted back. Manfred recognized the scene. Kurt did not like to be pressed. He liked air around him. He needed room. Manfred had been through it often enough, waiting on line with Kurt at the movies, crushed in at soccer games, standing together in the rush hour on Frankfurt's trolleys. Crowding made his brother crazy. Kurt's black eyebrows arched in indignation. The woman turned away. Kurt faced front again, his eyes squinting in anger. All morning he had been squinting and breathing in a powerful rage.

"Shit," Hugo Bondi said, alongside Manfred. He too was breathing irregularly, gasping at the cold saline air as though it might save his life. He was at least two inches taller than the rest of the boys. Manfred glanced up at him, glanced back at Kurt. "You'll get terrible gas pains doing that," Manfred said. Hugo Bondi ignored him. He began to belch in great rasping sounds. Manfred recognized those, too; Joachim Adler was a master at them. "Listen," he said, turning to Scranton Boy, who was now holding onto Manfred's elbow. He didn't know what he was going to say. "Your name?"

"My name."

"You have a name?"

"Erwin Marks."

"That's your name?"

"I just told you."

They shook hands. Erwin tried to hold on.

"And you?"

"Me?"

"What is it?"

"Manfred."

And they shook hands again, feeling foolish.

"We're moving," one of the Heller brothers whispered to the other. "You feel it?" The ship was actually slipping away from the dock, sidling slowly toward midstream. A tug struggled at the stern, another hovered midship. The twenty feet that separated the *Europa* from the dock widened to thirty, then forty. Where was Kurt? Manfred looked for a black overcoat among many black overcoats, then for a unique white scarf. "I can't find my brother," he said. "There," Erwin Marks said, shading his eyes. "There in front of my mother and father." They were moving slowly downstream. The ship shuddered as it pointed toward the sea, straining against the unnatural, slow pace. There was a cheer from the upper deck, a final burst of confetti. Thin streamers of weightless paper, pink, pistachio, green, white, yellow, even silver, floated in front of their eyes. Kurt was waving his arms overhead. He was no longer squinting. Mr. and Mrs. Marks huddled together. Erwin's father chewed his lips, while his mother kept pointing at her son. She was pantomiming a message. "What does she want, for God's sake?" Erwin said. Her arms flailed enthusiastically in the air. "Why are they doing this to me?" Erwin Marks asked. "It's stupid. They're stupid. They don't even know where Scranton is."

"Goddamned fools," Hugo Bondi then said, in a slow drawl. "All of them. And good riddance, good riddance to the whole shit-filled Reich."

Kurt clasped his hands over his head in the symbol of victory. Manfred made the gesture, too, so that Kurt would know that he had seen him. While they acknowledged each other, the ship began to slide forward, picking up speed. Manfred lost Kurt then, found him a moment later, held him there a second or two, lost him again. He kept staring back at the dock, forcing himself to concentrate, to hold on, but it was all slipping away, it was going. The ship ground ahead. Sky and water flooded the way. Below Manfred, a small packet-boat paralleled the *Europa*, piloted by a sailor wearing a yellow slicker. The sun, pillowed by vast soft clouds, was already high. Soon a white radiance spread east and west, over the cloudbank. Manfred forgot where he was for a moment, his face emptied, he was suddenly bewitched by the new double magic of space and seascape. He breathed it in, tasted the salt on his tongue. The wind flicked his cheeks like grains of sand. The horizon lay at the end of the world. Back there, the wooden dock was all one face now. Kurt was gone.

Manfred let go of the railing, turned away aimlessly for a few seconds, turned back without speaking. "I don't feel good," Erwin Marks said. As Erwin spoke, the Heller brothers disappeared below deck, keeping a dignified silence. Their fine red hair flew in the wind, their brown freckles were like raisins. Brothers, Manfred thought . . .

"The fleet's in," Erwin Marks suddenly shouted, with new spirit. Manfred snapped to. Leaning out, he saw several auxiliary naval vessels floating at anchor a little way off. The shoreline beyond them was dotted with fat oil tanks. As the *Europa* moved downstream, three cruisers came into full view, lined up one behind the other, parallel to the shore. From a mast of the first, a sailor waved them on. Manfred waved back. A couple of crew members were polishing the guns in the stern. The second vessel was silent, its artillery cased in canvas. A party of officers strolled the deck, oblivious to the *Europa*. The *Europa* gently tooted its horn. On the third cruiser, part of the ship's complement stood on deck, facing them, lined up in a row in perfect order. They wore white uniforms, their snowy silhouettes set rigidly against the great flat coastline and the late morning sun. *Emden*, it read on the ship's hull. Manfred stared at the *Emden*'s crew. That was what Julius Vogel had looked like once, eighteen years ago, before he lost his middle finger. Legs set slightly apart, hands folded behind his back, naval cap pulled smartly down over his forehead, standing at ease in the service of the Fatherland. In the wind, the swastika blew on the *Emden*'s stern. Little pennants, each a different color, fluttered from the mastlines. A band on board the *Emden* began to play marches, ancient *fanfaren* tunes that dated from the time of Frederick the Great. Stirring trumpet sounds curled sharply across the water as they sailed by. The drummer-boy thumped on two kettledrums, holding the drumsticks high over his head between each beat. While the *Emden* crew saluted from their deck positions, Hugo Bondi marched in place, in time with the music. Erwin Marks joined in. Some of the other passengers applauded. "Buggering heroes," Hugo Bondi shouted, still keeping time. "Forty-seven shitholes lined in a row. I'll give them a taste of life at sea, if that's what they want. Take that," he ordered, farting as he highstepped at the rail. The boys began to laugh. "And that," Hugo Bondi added, farting again. But this time there was also a sudden roar from the *Emden*'s own stacks, which let go three vast bellows of farewell that shook the straits and deafened Manfred and his new friends and all the other passengers on the *Europa*.

They were finally on their way out to sea.

The afternoon passed; the sun vanished; thick grey clouds chased them into the North Sea. A routine immediately began to establish itself, enforced by the crew and their officers. Everyone paid attention. The boys had lunch at 12:15. At 4:00, if they wished, they could have tea and dry biscuits. Their dinner was served at 6:30 in the third-class dining room. Bells rang regularly. Steward Leutze, a sign read, was on duty in a cubbyhole near their cabin, but by dinner he had still not appeared. The Heller brothers tailed each other all day, never letting the other out of sight; everything between them was still a secret, whispered into the other's ear. Together, after tea, they went out into the corridor and shared a bar of chocolate. That left Manfred, Erwin, and Hugo Bondi, sitting idly, side by side. Hugo Bondi was a heavy-boned young man, already thick in the shoulders, with a large slanting jaw and hair sprouting maturely from the base of his throat. "How old are you?"

Manfred wanted to know, playing with his teacup. "Older than you pansies," Hugo answered, matter-of-factly. "We're not pansies," Manfred said. "Come on, how old are you?" "I'm seventeen," Hugo Bondi answered, expanding his chest. "And"—breathing deeply—"I've had ass every week of my life since my fifteenth birthday." A tremor ran through Manfred at this news; Erwin Marks giggled; and Hugo Bondi looked nostalgically out to sea, over the tea-things, and gravely touched his penis.

Later in the day, Manfred and Erwin made a tour of the ship together. Whole areas were forbidden to them, entire sections roped off with forbidding signs that excluded entry to third-class passengers. They caught a glimpse of huge steamer trunks delivered to larger cabins on the upper decks. A woman in a fur coat, carefully posed, was being photographed outside the ship's library. In the chrome bar, a gentleman in golf knickers sat drinking Scotch whiskey out of a tall bottle. Someone was already playing the piano on a small bandstand. A map on the wall traced their course. A blue pin was stuck about half an inch from Bremerhaven. They examined a diagram of the ship tacked on the bulletin board, checked the weather report alongside it, and memorized the history of the *Europa*. Six clocks hung on the wall, each giving the time somewhere else in the world. Already it was clear to Manfred that he loved life at sea. It was ordered, it was busy; yet it left him free. It was also circumscribed, constrained; but the whole world seemed open to him. In precisely one hour, no more, no less, he would have dinner with his friends. No one would care whether he showed up or not; no one would care how much or how little he ate. He already had his life-saving drill instructions, he knew where his station and his life jacket were, he was ready for all possibilities. Sailing at his side, he had a faithful pilot fish, for whom he already felt responsible, a frail but firm new friend who made dramatic threats that he never carried out. Erwin Marks would faint at any cause, he would vomit, he would die. Yet, in his excitement, he outran Manfred on their tour around the ship. "Come *on*," he urged, as Manfred stopped to examine still another clock, another map, another piece of ticker tape, mysteriously flowing out of a machine with the news of the day printed on it. Afterwards, Manfred lay exhausted on his bunk, going through his new Cassell's. Thin paper, which he delicately fingered, ink of varying shades to indicate the endless classifications of language, a section each for proper names, slang, and abbreviations, and thousands of beloved words, anchors to reality, half of them English. Kurt's choice and a perfect one. Re-mote, Manfred read, mouthing the word silently. Rep-li-cate. Re-ten-tive. Re-tract. Ret-ro-grade. He had to check each one. He had to learn their meanings. The *r*'s were hard to say in English. Re-ver-ber-ate, he read, hesitantly. Re-vet-ment. He would memorize the whole book.

Across the cabin, Erwin Marks babbled on, peering out from between Hugo Bondi's legs, which dangled from the bunk above. "I always wanted to play soccer," he was saying, "but I suffer from shortness of breath and a tendency to asthma. Have you ever had asthma?" He couldn't quite get the *s* right. "Athma" was what they heard. "Have you?" he asked again. No one answered. "One attack can kill you. It keeps your breath inside, you could strangle. It has always prevented me from fulfilling my dreams of athletic achievement." There was a skeptical snort from the upper bunk.

Re-vile, re-volt, rhi-ni-tis, Manfred read, checking each meaning in detail, then cross-referencing it with the German section. He had a lot of work to do. Adele Gordon had warned him that English was full of dirty tricks.

"I tried to fence, too," Erwin went on, undiscouraged. "It was my dream to be invited to join the Rhenania Student Fencing Club. I actually did fence for a short period with that goal in mind, but my mother, my father, too, they were both afraid that I might be scarred for life."

"Scared for life is probably more like it," Hugo Bondi remarked from above, looking down sardonically at Erwin.

After dinner, which turned out to be the same kind of Baltic fish Manfred had shared with Kurt the night before, the three friends took a stroll on deck. The Hellers were off on their own. Outside it was cold and wet. The boys faced ahead, eyes half-closed against the mist. They could hardly see where they were going. Space had disappeared with darkness, the seascape was now only a sound that traveled with them on each side of the ship, hissing. To the northwest, Manfred could make out a cluster of lights, which soon disappeared as the ship rose and fell. With each dip, Erwin reached for Manfred's elbow.

"You guys know that you're perfect German types?" Hugo Bondi said, into the wind.

"You'll have to talk louder."

"I said, you guys are perfect German types."

"What do you mean, perfect German types?" Manfred asked.

"There are six perfect German types," Hugo Bondi shouted. "Marks here is a Rhinelander type and you're Nordic."

"Where'd you get that?"

"I just know."

"If I'm Nordic, then what are you?"

"I'm probably Baltic. With me, it's hard to tell. I'm not a perfect type. I have a big frame. Everything about me is big." In the dark, he touched his penis again. "Not everybody is a perfect type," Hugo Bondi continued. "Christ only knows what redheads are."

"You don't believe in that stuff?" Erwin Marks asked, striding along on the other side of Manfred.

"It says."

"Says where?"

"It's a proved fact. It's not true for everybody, of course. But for enough. There are six basics. I'm probably Baltic, I'm so big, and something else. Maybe the redheads are Dinarics."

"Proved by who?" Manfred asked.

"Everybody knows it."

"How can a Jew be a Nordic?"

"For Christ sake, don't you guys know anything?"

"The Nazis say we're not even German."

"The Nazis say anything. Don't fall for that. They're shitholes. What color are your eyes?"

"Blue."

"I could've told you. You're Nordic. You have to be Nordic. All Nordics have blue eyes. Now look at Marks's black hair and the rest. That's Rhineland. Rhineland is black hair. And I'm Baltic. We're perfect German types. Or you are. I'm maybe only half."

By now, they were at the prow of the ship. The deck slanted upwards, towards the knife-edge. A crew member stood alone at the point, wearing a sodden woolen cap and a pea jacket with the collar up. The wind cut their faces now, the mist felt like shattered glass. Behind them, they could see the lights of the *Europa*, around them strange funnels leading below deck, above, through a long window streaming water, an officer on watch behind the wheel. In the dark, Manfred laughed to himself. Nordic. Nordic and Rhineland. Erwin Marks did have black hair, but what did that prove? Half the world had black hair. And Dinaric. Manfred had never even heard the word before. Hugo Bondi probably made it up. He was the kind who was always making things up.

"We'd better get our asses back," Hugo Bondi said, "before we drown."

They started back then, arm in arm. "You guys know how to do it?" Hugo Bondi asked, still shouting.

"Do what?"

"What do you think?"

"You tell me."

"You must be virgins."

There was no answer.

"You are virgins, aren't you?"

Again, they were silent.

"Pay attention now, because I'm only going to say it once. You guys are lucky you found me. Nobody ever had to tell me, with me it was just natural. I always knew. First thing, you just tell them you want to touch it with yours." There was a pause. They had reached the steep metal stairway that led below, were on their way down sideways. "That's it?" Manfred asked, after a moment. "Jesus Christ," Hugo Bondi said, looking up at them from the bottom step. "Now pay attention. Naturally, you're on top. You know that much, don't you?" They proceeded into the maze below deck. "You make sure you tell them you won't do anything else, just a touch. You swear it. Then you touch it with yours and it gets wet and you wait a few minutes, that's the hard part, you have to wait, you tell them you need them, you're in terrible pain, it hurts. Then you touch some more, a half-inch maybe, it gets wet, and again, it's easy, you don't have to do a thing, it happens." They were almost at their cabin now. "Just swear at first you won't do anything, that's the important part. You guys listening? Nobody knows better than me. You don't know how lucky you are. I know what I'm talking about. I've been laid a hundred and nineteen times since I was fifteen and maybe only three times with a whore."

The rest of the evening was ordinary. Hugo Bondi lay in his upper bunk and described to them the various modes of contraception popular in the Orient. Some of them had to be explained to Erwin Marks twice, they were so amazing. The Hellers returned from the smoking room, where they had sneaked a cigarillo apiece, and sat down on the floor of the cabin to play cards, arguing quietly over each point. A few minutes later, Samuel Heller got up and solemnly passed a sourball candy to each of their companions. "For you," Samuel said to Manfred, holding out a little lozenge, wrapped in silver paper, like Uncle Selig's. "For you,"—to Erwin. "And for you,"—to Hugo Bondi, who liked candy. In his bunk, Erwin was busy writing on a pad of lined paper. His first letter home, he told Manfred. He had promised to write every day. His mother had made the same promise. They would write to each other every day for the rest of their lives. Erwin wanted Manfred to have a look, he said, a little slyly, before he sealed it. For spelling and that sort of thing, he added. However, he didn't want to impose. *Bitte* . . .

"Whenever you're ready," Manfred said.

"I'm ready," Erwin said, thrusting the pad at Manfred.

"Dear Parents," Manfred read, with a touch of guilt. It had not even occurred to him to write to his own mother and father. "On my way," it said, "I have fainted because I am homesick. Dear parents I want so much to come home otherwise I'll die of a broken heart only because I am very homesick for all of you dears. Dear parents please try to get me home soon, not a 100 horses can keep me here because I am homesick. I am even homesick for my Dear Dear brother Gert. Even at night I cannot sleep because I am homesick. The best is that you send me a telegram, that is the quickest way. I cannot stand it for homesickness. *Homesick*. Be greeted and kissed by your dear homesick Erwin. *Homesick*."

"What do you think you're doing?" Manfred asked, after a moment. He could feel his heart beating in agitation.

"What do you mean?"

"You can't send this to your parents."

"Why not?"

"Because it's all lies. Because it'll make them sick."

"It's not all lies," Erwin said. "It's how I feel."

"You guys better speak in English," Hugo Bondi said from his upper bunk. "You sure as hell can use the practice."

"Tear it up," Manfred said to Erwin Marks, "and start over again tomorrow. You just said good-bye to them. How can you write something like this? How can you do this to them?"

"Let me have a look," Hugo Bondi said, making a halfhearted swipe at the letter, but Manfred held on.

"This will make them sick," he said again, in a stern headmaster's voice. "You know that."

"They sent me away."

"For your own good."

"Jesus Christ," Hugo Bondi said. "Cut the shit."

"Oh," Erwin Marks moaned, rolling over on his stomach.

"Now what's wrong?"

"I have a headache. I have a stomachache. I'm going to throw up. I want to die."

"God forbid," Samuel Heller said, looking up blandly as he trumped his brother in the middle of the floor.

"I'm going to sleep," Louis Heller mumbled.

"I bet none of you guys ever fucked an animal," Hugo Bondi said, sitting up alertly.

The ship rose and fell, the prow smashed into the waves.

"Please don't vomit in here," Samuel Heller said to Erwin Marks. He popped another sourball into his mouth.

"You're not going to mail this letter, are you?" Manfred asked.

"Give it to me," Erwin said. "It's mine."

"I must tell you all now that we get up early in the morning, Louis and I," Samuel Heller said, in rabbinical tones. He began to shuffle the cards. "Because we put on phylacteries every day and say the morning prayers."

"Phylacteries?" Hugo Bondi asked, after a moment's silence.

"Didn't the Nazis tell you that my brother and I are Jews?"

"Very funny."

"We put on our phylacteries every morning," Samuel said, still shuffling the cards. "We never miss. And we don't make love to animals. We're pious."

"You getting smart with me?"

"They say you can't be a success in America if you're orthodox," Erwin said halfheartedly. He got up from his bunk. "They say you have to live the way everybody else lives."

"They say that about everywhere," Samuel Heller answered.

"Jesus, Orthodox," Hugo Bondi said.

Erwin reached for Manfred's elbow, for a touch. He stood there, in the middle of the cabin, holding on to his new friend. They smiled wanly at each other, Rhinelander and Nordic. Erwin made a puking sound. "No,"

Manfred said, reaching into his pocket for Dr. Schwabacher's tiny blue vial. "Now hold on," he ordered, "and watch me." He opened the bottle and carefully tipped a drop of the fluid onto the end of his index finger, then put it into his mouth and sucked it in. There was a faint, not unpleasant odor; the taste filled his whole throat. It was very smooth. "What's that?" Hugo Bondi asked. "It's something for stomach trouble," Manfred answered. Then he offered the bottle to Erwin, who, without hesitation, also measured out a drop and swallowed it.

"Again," Erwin said. They shared a second drop, and a third. Finally, after a fourth, Erwin asked, "What is this stuff?" "Opium," Manfred said in a loud voice, wanting to astonish them all. Then the door to the cabin opened. A face appeared, lighted from below by a flashlight throwing off a red beam. The hairy nostrils were distended, the bloodshot eyes opened exaggeratedly. Two fingers pulled the mouth down at either side in an effort to achieve a spectral effect. "Opium," Erwin Marks shouted at the apparition and fell to the floor.

"So these are my nice Jewish boys," the face said, shining the flashlight on Erwin. "Another batch. My name is Klaus Leutze and I'm your steward," he added with a crooked smile. "Anything I can do for you? What's wrong with him?" Erwin was already scrambling to his feet. The Heller brothers also got up from their card game. "I'm here to serve," Steward Leutze said, in a sincere voice, having been generously tipped and stuffed with special instructions by Kurt Vogel and Erwin Marks's parents before they left the ship. "You must be the one with the vertigo," Steward Leutze said, peering into the cabin. "Herr Marks, I presume?"

The long night began. The ship sailed north-northwest. A storm followed them from the Danish coast, heavy rain and strange, early-winter thunder. A whole new rhythm had overtaken them and already they had begun to move and breathe unthinkingly to its demands. Even Erwin Marks's delicate body was responding evenly, without thought. The five young men turned into their bunks. Sighs multiplied throughout the cabin, there was a hacking cough from one of the Hellers. After the single bulb was turned off, Hugo Bondi made a dutiful attempt to describe the long-term effects of untreated venereal disease, but was shouted down by the others. It was too late in the day, their overstuffed minds refused the information.

Hugo was the first one asleep. The pious Heller brothers followed him, then Erwin Marks.

A half-hour later, still awake, Manfred got up to use the public toilet down the corridor. He stood in front of the strange tile wall in his felt slippers and urinated sleepily onto chunks of ice. When he was done, a few drops of urine fell onto his pajama pants. Next door, sitting straight up in his closet, asleep, Steward Leutze was snoring. Manfred tiptoed past. Back in his bunk, his head began to swirl as soon as he lay down again. It was not quite dizziness, it was not at all like being drunk in Bremerhaven. This time, there was a nice, curving rhythm to it, swinging him back and forth gently on its own momentum. Oh, yes, he thought dispassionately, arching higher and higher in his bunk towards the welcoming white radiance that stretched overhead from one horizon to the other. A few sweet seconds passed as he heaved his body up, then the rhythm hesitated, the pure white welcoming heaven vanished. Manfred moaned and rolled over onto his stomach. A kind of heaviness began to set in. It was very thick. Oh, yes, he thought again, lying there face down on his bunk, inside the heaviness. He moaned again, into his pillow. The heaviness encased him. He was bound by it, unable to move. Manfred's eyeballs rolled back in his head, he felt sick to his stomach, his tongue flapped inside his mouth. Slowly sobriety was returning, with a giant's tread. His tongue flapped again. He desperately needed a drink of water. He was suddenly parched. It was very strange.

There was a steady snore in the cabin, a couple of coughs. A few feet away, in a lower bunk, Erwin Marks twitched in his half-sleep, mildly hallucinating as the *Europa* finally sailed out of German territorial waters. Manfred reached for his money-belt, beneath him. It was like trying to sleep with a life preserver around his waist. All the boys wore money-belts. Manfred's was fat with bills. He could feel that they were still there, safely tucked into the canvas pockets. England's coast, he suddenly remembered, lay to the west. Benno Mann, the family pederast, was an Anglophile. So was Lotte Mann, whose favorite writer was George Eliot. Kurt Vogel was going to England soon. It would be his new homeland. Kurt Vogel had green eyes. Maybe he was a perfect Dinaric, like the Hellers. The Swedes were also somewhere around. Where the Swedes lay, there lay frigid Norway as well. The North Sea was dense, water was thick. Manfred could feel the resistance of it through the hull. It was like the heaviness in which he lay, paralyzed, waiting for someone to bring him his fill of water to drink.

Forty feet from where he slept, in the stern, four massive pistons churned up and down with an echoing metallic sound that never ended. Behind them the giant screw turned powerfully. Somebody's body twitched in one of the bunks. A leg shot out in a muscle spasm. The snoring was passionate. Down the corridor, Steward Leutze was laboriously getting to his feet to make his midnight rounds.

Kurt, Manfred thought, in the last echo of his opium dream, mindlessly invoking the magic names of the gods at last. Kurt.

It was past midnight as Steward Leutze padded by the boys' cabin, flashlight in hand. It was already the next day. A white scarf fluttered in the wind. Confetti fell through the air. A terrible thirst sandpapered the roof of Manfred's mouth. Kurt, he said aloud. *Vater. Mutter.*

Part Two

Chapter One

It was like falling out of the wrong end of a telescope; or awakening from a forty-year sleep; or spiraling nose-down in a plane out-of-control, rushing head-on to meet the spinning landscape below.

He had been trying to hold the new world inside his head, stuffed there in miniature, shrunk to human brain–size, standard repeated images that he had come to take for granted in Frankfurt-am-Main of New York, Baltimore, America. Now it suddenly emerged as itself, beneath a glaring sun, a stony manufactured force of human nature beyond nature, a swollen manmade island resting threateningly to the east, in isolation. There was nothing commonplace there, nothing ordinary; it went on for miles, into infinity. It could mean the end of Manfred Vogel, to be lost out there alone for life, address unknown, identity splintered into countless shards. The clarity of the light was no help. He had never seen such sharp lines, such jagged, mean angles. Everything stood brutally plain from the ship. There were no softening curves, no folds to hide in. In their place was a dense straight thrust upward, massed as far as his eye could see, soon separating itself into gleaming individual buildings, standing crystal-like alongside each other in the brilliant American sun. He made his hands into a kind of binoculars through which he stared at Manhattan. Fingers of piled wood thrust out into the river. Nothing moved. There seemed to be no shadows, no signs of human life. Human life had disappeared, only stone existed, a glittering stillness wherever he looked. He began to count the buildings, to have something to do, at the count of three recognized the Empire State; everyone around him recognized the Empire State at the same moment; a murmur of pleasure spread among the passengers who lined the decks. A little behind schedule, the *Europa* shuddered upstream. It was late afternoon.

At the Hudson River dock, Max Gordon rushed Manfred through the bureaucratic swarm of customs, immigration, and agency social workers.

They all wanted something from him, papers, money, fingerprints, handwriting samples. Sign here, he was ordered. And here, on both lines. Open the big bag. Any food in there, anything edible, anything *alive?* Just sign that, right there. How come your papers say Leopold and you say something else? My name's not Leopold, it's Manfred, I told them in Germany. He did, Max Gordon said, shoving ahead. I have proof, it's Manfred. Look . . . Max pushed on, solving the confusion. He was alone. He had left his wife and daughter in Baltimore. He was very efficient, he made sure that Manfred moved along. But he was nervous, too; he spoke too loudly, in unnatural bursts that seemed to take him by surprise. Say good-bye to your friends, he said. Where were his friends? Over his shoulder, Manfred caught a glimpse of Erwin Marks, seated in the backseat of a red automobile, driving off with his foster-parents and his two new brothers, on their way to Scranton, Pennsylvania. The others had disappeared. Hugo Bondi was gone. So were the Heller brothers. The social worker said that a wire was already on its way to the Mendelssohnstrasse, confirming safe arrival, health, and good spirits. That's fine, Max said, glancing at Manfred. It'll save me the trouble. Manfred smiled wanly. Are you in good spirits, Max asked. He eyed Manfred with both eyebrows raised, looking surprised again. Don't mind me, he said, I'm just being facetious.

Soon Max Gordon and his ward were on their way to the New Yorker Hotel, on Thirty-fourth Street. The lobby inside was as big as the Schillerplatz. Max already had the key to their room. A little man wearing a short, tight jacket that was missing a button carried Manfred's bag upstairs. The elevator operator stopped at the eighteenth floor to let them off. They stayed overnight, sharing twin beds in room 1806. In the bathroom, there was a shower that ran hot and cold at Manfred's touch. Outside the double window in their room, down below, huge letters hung in the air. S-C-H-E-N-L-E-Y, Manfred made out in white bulbs that blinked on and off. Others ran pink and blue. M-C-A-L-P-I-N. H-O-R-N-&-H-A-R-D-A-R-T. A yellow glow wavered along the street in a slow, watery line. "You might as well take it in while you can," Max said, eyeing him sympathetically from his bed. "They don't build them like this in Baltimore."

They went across the street for dinner to the Hotel Pennsylvania. A dance band played while they ate. Manfred had lamb chops. The dining room was half-empty. "You like this place?" Max asked. "I thought it would be a treat your first night in America. You like this kind of music?" Then ten minutes later: "That your stomach grumbling? Want a Bromo? What'd they feed you on that ship, anyway?" Another ten minutes: "Don't make any snap judgments. You'll get fixed on some crazy idea and it'll take you years to get rid of it. You don't have to assimilate everything in one swoop." Then: "I don't mean to keep nagging you, but wait until you've been here at least three days before you jump to any conclusions. Don't evaluate anything, me included, all the Gordons included. You're already judging me, aren't you? I can tell by that look." Max laughed. "Am I speaking too fast for you? You sure you understand my English?" Max ran his hand through his white hair. A ripple of waves spread above each ear. His skin was the color of honey, and he smiled a lot. "Finish your chops now," he went on. "Don't pick at them. Florence doesn't like that." All through dinner, the band played American tunes while a lady sang through her nose into a microphone. A few couples were dancing on the shining floor. Soft pastel lights hovered over them, baby blue, pink, flattering off-white. They were not like the lights Manfred had seen from the hotel room, which opened everything up like a surgical cut. Manfred finished his dinner with ice cream after watching the dancers fox-trot between courses. While Max paid the check, a black man took away their dishes. Then they walked around the block to get some air. Along the way, Max bought a pack of Chesterfields at a candy store. As soon as he finished one cigarette, he lit another. "By the by," he said a few minutes later, undressing full-out in their room. "You don't have to tell anybody that I've been smoking so much. I usually have more control. You'll be doing me a favor." He winked as he slipped out of his BVD's and into pajamas. Manfred nodded silently. Max's body was almost hairless, with a small black bush standing out at the crotch and a puff of grey curls on his chest. He also had a small curving penis. Manfred had never seen his own father naked. He was blushing. His brain still seemed to be in the hotel elevator.

The trip the next day took six hours. They left from Pennsylvania Station, down the block. A plume of black smoke accompanied the train all the way to Baltimore, coursing along the left side of the track, where the wind took it. At North Philadelphia, a vendor boarded the train and sold Manfred and Max two Hershey bars. They also had chicken salad sandwiches that Max had bought at the station for the trip. On the way, they shared the *New York Times* and chatted perfunctorily, as though they were old friends who had long ago said everything important to each other. "Why don't you call me Uncle Max?" Max suggested at one point, sounding relaxed. "Would that make it easier?"

Around Wilmington, Manfred took out his little address book and began to flip through the pages. He already had an American section. Heller in Cleveland, Ohio. Bondi in Louisville, Kentucky. Marks in Scranton, Pennsylvania. In care of Wise, Friedman, and Schoenberg, respectively. Before they docked the afternoon before, even before they had sailed past the Statue of Liberty, Erwin Marks had already vomited on deck. "Oh God," he kept saying, looking at the mess at his feet. He still held on to Manfred's elbow, shaking with excitement. The night before their arrival, Hugo Bondi had finally taken a bath after five unwashed days—Steward Leutze ran the water himself, holding his nose—then shaved carefully and repacked his slovenly bag. It was filled with dirty laundry, but this time nothing hung out of the corners. Oh, the acrid odor that came from Hugo Bondi's bunk, the smell of ancient piss in the windowless cabin, of crusted semen. Through it all, the Heller brothers wandered as though they were going to be anointed as a single holy being. The two redheads seemed to rise in the air together at the first sight of American soil, they appeared to be levitating with irresistible spiritual ecstasy. Louis's hands, Manfred noticed, trembled passionately as they held onto the ship's railing. Manfred himself had stomach cramps. They doubled him up and made him sweat as the buildings to the east came into sight. Still, he decided against opening up the tiny blue vial. The sweet scent of opium would give him away. Max Gordon would think he was some kind of dope fiend. He felt like a fiend, thinking about it.

Were his ears clean? They had a tendency to accumulate wax. Why wouldn't his hair stay down in the back? Two cowlicks stuck up maliciously. Manfred poured water on his hair until it was slick. Then he made it even slicker with his hairbrush. Heading down the gangway, he kept his eye out for a head of white hair. The Heller brothers were already standing on the dock, a pair of anxious-looking adults holding them by the shoulders. Manfred had one last cramp, then stood up straight. On his lapel, he wore a neat little card that read: "My name is Manfred Vogel (Frankfurt-am-Main)."

When they got to Baltimore, Max Gordon left Manfred standing alone in front of the station while he went to get the car. He returned in five minutes driving an old battleship-grey Buick. The knob on the gearshift looked like it was made of real marble. It was colored in beautiful purple-and-white swirls. "Hop in," Max called, after stowing Manfred's bags in the trunk. Manfred sat up front, next to Max, against a woven backrest. "Got everything now?" As he spoke, Max pulled down on the beautiful gearshift as if it was an enemy, then thrust it forward punishingly. Manfred heard a horrible grinding noise, Max pressed down on the accelerator, and the car shot off. Manfred's head bounced off the back of the seat. This happened at every light. "Sorry," Max said, a little shamefaced. "You OK?" Manfred looked straight ahead, through the windshield. There was nothing to say. There was only scenery. A new city flew past, softer landscapes, a million blurred bricks. Stone piled on stone, houses sat on steep terraces. Before they had traveled a couple of miles, the Buick had stalled twice. "I don't know what's got into me today," Max said, gripping the shift.

They drove on in silence, heading north. The passing freedom between them had vanished during the train trip. They had deposited it at the New Yorker Hotel with their room key when they checked out. Now they were on to something else, there were powerful new terms to deal with in Baltimore. Manfred didn't trust himself to speak, he felt so exposed. It was like seeing Max naked the night before and this morning, too, after his shower, rubbing himself down with a heavy towel. Max lighted his second cigarette, missing the tiny tray on the dashboard and spilling ash all over the floor. "Baltimore," he breathed, nervously. "Wait'll you see it in the spring."

They were driving alongside a park, past a dim lake enclosed by a metal fence. Max waved toward it vaguely, then grabbed the wheel with both hands and passed a car in front of them without signalling. The driver of the car yelled something, Max jerked forward, Manfred's head bounced against the back of the seat again. By now he was sitting at rigid attention—sphincter tight, toes stretched, eyes bulging—as he had trained himself to do in Frankfurt in the presence of madmen. "We call it Druid Hill Park," his friend was saying, in his loud unnatural voice. Again he waved one hand, this time in the direction of a vast hothouse made entirely of glass. "Getting closer," he said. A spot of ash fell in Manfred's lap.

They made a few turns. The city streets broadened and became suburban. Max began to calm down. There were trees everywhere. There were rowhouses. Then there were grander houses in many styles. A yellow trolley ran in the middle of the street, on its own roadbed. Manfred liked a pink house they passed, with a Spanish tile roof and wrought-iron gates everywhere. There were other houses he liked. One had white pillars in the front and floor-length windows facing the street. "Just wait until spring," Max breathed again. Then the yellow trolley passed them, going forty miles an hour. Max blew his horn. The trolley sped down the track, its tail swaying from side to side, wires shuddering overhead.

Soon they turned off the wide street onto another wide street. The car rumbled over new trolley tracks, stopped for a light, stalled. Max sighed. Manfred curled his toes. Then they were on a small feeder street. The street was lined with rowhouses. Each house had a porch. All the houses looked alike. Some boys were playing ball in the street. Most of them wore long pants. Manfred thought of his German knickers. He twisted in the front seat, trying to get a good look at the boys. "Home sweet home," Max called, purring out of the side of his mouth.

"Pardon?"

"I mean we're here."

The boys made way for the car reluctantly, pulling slowly aside to the curb. One boy had white hair, not as white as Max's, more a cornflower yellow, all in damp curls flattened to his head. The boys stared at the car. They were wearing sweaters with strange geometric designs on them. A figure detached itself from the group and ran up the steps of one of the houses. "There goes Adele," Max said, shaking his head. "She's the quarterback. She calls the signals." Max parked the car just as Manfred heard a screen door slam on the porch. "Hey," one of the boys called. "Hey," another one said. With a half-smile, Manfred raised an index finger. A few heads nodded. One boy stood on his head against a tree, grinning upside down. Manfred sat there for a few seconds, listening to Max grind the gearshift into neutral, and tentatively wiggled his finger. When he got out of the car, he saw three balloons hanging from the ceiling of the Gordons' porch, gently floating in the fall breeze. One was red, another white, the third was blue. A vague sense of excited pleasure washed over him. He looked over his shoulder at the boys on the other side of the street. "Maybe I'll catch you tomorrow," the boy with the yellow hair said. Manfred nodded. A mess of sycamore balls lay on the pavement where he stood. He had to step carefully to avoid crushing them as he picked up his bags. "Here, I'll give you a hand," Max said, grabbing the canvas sling. The boys stared at him and tossed a football back and forth without looking at it. Six of them were in long pants, two in knickers. Manfred touched his wallet, touched his money-belt, reached for himself between his legs. His hands were shaking. "Come on," Max called. "The ladies are waiting."

Aunt Florence had thin bony legs that were slightly bowed. Aunt Lotte had legs like that, almost without flesh. Aunt Florence looked a little like Gloria Swanson, with certain distortions: longer nose, quicker eyes, a mouth somewhat less full, less generous, but her smile curled downwards in the same inviting way. "Let me take a good look at you," she said, pulling Manfred towards her in the living room. She held his cheeks between her hands and looked steadily into his eyes. Florence Gordon and Manfred Vogel, like Dr. S., were the same height. "Blue eyes," she said, her smile curling downwards now, "with a little green in them. I bet they change color with the clothes you wear. I told everybody what you would look like. I had a feeling. I said blue eyes. And skinny. We'll fatten you up. Do you wear braces? That front tooth . . . " Her hands gripped his shoulders. "Skinny," she repeated, feeling his bones. "Well, what you don't see you ask for. Everything in this house is yours, as though you were born to it. Can you believe it, he's here. It's happened." She put a handkerchief to her nose. "Don't worry, you'll never see me cry again. It takes a lot to make me cry. I am not a weeper."

Behind her mother, Adele—full-bodied and full-faced, not at all skinny, no bones in sight—stood with one foot pigeon-toed in and stared at the floor. Her skin was white, without highlights, no pink, no flush, no color at all, her hair a kind of coppery tone, hanging straight down to her neck, not frizzy, as the drawing in her letter suggested. She had a dirt streak on her chin, in the little half-moon cleft, and a serious look on her face. In a corner of the room, Max cleared his throat. "It's over," Florence said. "I told you, I never cry." Max cleared his throat again. "They're worried I'm going to embarrass you," Florence went on. "They made me promise, no fuss. Fuss makes them nervous. You don't eat salmon, do you?" Manfred looked surprised. "I know about all those things," Florence said. "Allergies. Hives. We'll just have to learn how to live without salmon."

"We never eat salmon around here anyway," Adele said, still staring at the floor.

"You don't have to give it a second thought," Florence continued. "Your mother and I . . . " Max cleared his throat a third time. The room became silent. Manfred gave one of his bags a reflexive kick. At last, Adele looked up and broke into an easy smile, full of startling white teeth. "Come on," she said, making a move for the door. "I'll show you your room. Before she talks you into the grave."

"I'll give you the grave," Florence said. "And you could use a washing-up, too. Look at you. *Schmutzig*, all over your face. Football!"

"OK, OK," Max called out.

"Come on," Adele said.

First she showed him the downstairs: small living room, where they were standing now, with a look of pained suspense on their faces, small dining room right behind it, moving on to a breakfast room and a tiny kitchen. The rooms were full of ordinary furniture, oak pieces, china figures, fringed lamps. "This is where I do my homework," Adele said, pointing to the enamel breakfast table. Florence was right behind them. "She's got a perfectly good desk in her room upstairs and night after night she's sprawled out down here," Florence said. "I don't know how she gets anything done, God forbid she should miss a trick. Anyway, the poor boy's not interested in where you do your homework," she added, as they wandered around together. "Why don't you help him upstairs with his bags? He must be exhausted. How about some cocoa before supper? You look like you could use a pick-me-up."

No, Manfred said; no, but thank you. He didn't know what he could use.

"OK, let's go," Adele said, smiling her open smile. "I haven't got all day, either." She wore a heavy purple sweater that was pulled down to her thighs and brown-and-white shoes that were smeared with dirt, like her chin, only there were heelprints on her shoes. Manfred couldn't keep his eyes off the large, beautiful teeth that filled her smile. Somehow, her letter had not prepared him for large, beautiful teeth. "Don't you even want to see your room?" she asked, as Manfred hesitated.

"Leave the boy alone, for God's sake," Max finally said, from his easy chair in the living room. "You'll drive him crazy, both of you. Don't pay any attention, Manfred. Take your time." He was already a new man in his own house, his voice was normal, he had found his dignity and his self-assurance again in the seat of his velvet chair. Now he picked up the *Evening Sun* and began to read it, starting with the last page; that was Max's way with the papers, to go from back to front.

Somebody called Adele's name from the street. "I can't come out now," she yelled from the screen door. "Play without me." Manfred stood alongside her. In front of the house, the football bounced off the roof of the Gordons' Buick. "Let's watch that stuff," Adele shouted. The game went on, the boys rushed down the street after the ball. "Don't forget to tell her you think the balloons are great," Adele muttered to Manfred. "They're her idea. She means well." Adele rolled her eyes at the porch ceiling. "You want these balloons out here all night?" she called to her mother, over her shoulder. "Half the air's leaked out already." There was another shout from the street. "Jesus Christ," Adele shouted back. "I already told you. I'll see you tomorrow." "A-*DELE*," Florence called. "Show some respect." "Jesus Christ," Adele breathed into Manfred's ear. He could smell her perspiration, she was standing so close to him.

Upstairs, after Manfred put his bags down in the hallway, Adele showed him around. You could almost touch each bedroom door from the center of the hall. In the front was Florence and Max's own room. It extended the entire width of the house, sixteen feet from one wall to the other, and held a big double bed. "The passion den of Fairfax Road," Adele said, panting hard with dramatic irony. On a night table was a copy of an encyclopedic book three inches thick called *Eugenics and Sex Relations*, and on the wall over the bed hung a picture of a pale nymphet with pink butterfly wings on her shoulders. "In this room," Adele said, as she led Manfred out, "you knock before you enter. You never know."

Manfred nodded solemnly.

"In my room too, for that matter," she went on, glancing at him over her shoulder. "I value privacy. You might as well know that from the start. Try to remember, OK?"

"OK," Manfred said, in a submissive voice.

The bathroom was just off the hall, Adele then told him, pointing. Behind them, a foot or two away, stood Adele's room, smelling of talc and bubble gum, and next to it, Manfred's skinny cell, with the one closet. He peered in. At the other end of the room, there was a small window and a door. "Through there's the upstairs porch," Adele said. "I wrote you about the porch. Go on in, for God's sake, and put your bags away."

"I guess that's about everything," she concluded a few minutes later, hanging a towel over the back of a straight chair in his room. "Well . . . " She pulled the word out mysteriously, making it last. Manfred waited for her to continue. They were both looking at the floor with severe expressions. "I guess I should be getting out of your way," she finally said. "You must be feeling weird. Are you feeling weird? How could you not feel weird?" Without committing himself, Manfred shrugged. "Your face is still dirty," he heard himself say. "You'll get used to it," she answered. Then Adele went downstairs, flashing her powerful teeth on the way.

Manfred swung his bags up on the bed. A little blond fellow in a wet raincoat stood tugging at an umbrella on the headboard. Then Manfred pulled unsuccessfully at the venetian blinds on the window and finally left them at a tilt. He peeked through the glass panes of the door and saw a tiny porch, with a tarpaper roof. Under the roof, he guessed, stood Florence's kitchen. He could hear her banging away at supper down there. Behind the house and its backyard, there was a tin garage and an alleyway running the length of the block and another tin garage next door, ramshackle structures with no room for maneuvering a car. Overhead he could see the first sliver of a cold moon, spreading haze over the backyards. A-mer-i-kha, he said to himself with a perverse bite. A-mer-i-kha. The Gordon house and its neighbors were made of brick. Brick front, stucco back, some of it peeling. Drainpipes led down the walls from shared roof-gutters, sickly grey-spotted ivy twined around them. Next door, an orange cat lay flat on the fence, stretching its spine. The cat wore a belled collar. Downstairs, the kitchen door slammed shut. The cat disappeared. A figure ran down the back steps of the Gordon house into the yard, carrying a bag in one hand. It was Adele. Manfred could see her white face, the white parts of her shoes, her coppery straight hair. When she reached the garbage can, she lifted the lid and dropped the bag in, like a bomb. Then she covered the can again, turned to face down the alley with the same motion, and, cupping her hands over her mouth, began to make strange sounds. Manfred could see her blow into her hands, cheeks puffing with the effort; it was hard work. In a moment, she stopped and waited, then repeated the sounds. In another moment, the same sound came back at her from down the alleyway, only a few houses off, returned in a huskier voice. The second sound lingered for a few seconds, while Adele and Manfred both stood listening to it in silence. Before it was over, Adele placed her fingertips to her mouth and, opening them, blew a kiss to the moon. Then she ran back into the kitchen, skipping like a child.

Manfred waited a moment, hoping for something more, but nothing else moved, nothing sounded in the alley. Under the moon sliver, the haze drifted slowly from yard to yard. Patches of frost lay here and there in thin strips. Adele was downstairs again, inside the house. He could hear her talking to her mother, in normal tones, sounding domestic and conventional. Manfred turned to his bags then and began to unpack his clothes. As he worked, he felt a sweet anticipation that was all the sweeter for being unexpected. For the moment, it helped to put reality behind him; for the moment, it wiped out Germany, the *Europa*, and all the rest. He put away eight sets of immaculate underwear. Then socks. Then Uncle Benno's elegant handkerchiefs, Aunt Jenny's polo tie, shirts, pajamas, two sweaters, his old-fashioned expensive knickers. He removed his money-belt. It had left a mark around his waist. Free then, he rested on the bed, surrounded by his own possessions, making sure not to dirty the bedspread with his shoes. The sound of Adele Gordon's voice echoed from the kitchen and he soon dozed off.

When he awoke a few minutes later, he dutifully reached for his Cassell's. He would practice the language before supper, rehearse a little for the family downstairs. They would expect it. Manfred was supposed to be "fluent" in English. All his papers said so. That was his boast. Yet, he'd understood only about half of what had been said to him in the past two days. The rest he had had to guess at, using a kind of intuitive deduction that didn't always work. Everyone hungrily swallowed words or dropped them out of the corners of their mouths, like bubbles. Sentences were inverted, syntax smeared, pronunciation an invention of the moment. It was not like Herr Levi's classroom, where diction was perfect and grammar always serious. Max was the most dangerous. He didn't sound like anyone else, he seemed to have a dialect of his own. Manfred looked up "facetious" now, working hard at the spelling, and was surprised to find that in German it meant "witty." Was that possible? "Facetious" was how Max Gordon had described himself. Then Cassell added "joking." Well, perhaps; Max Gordon had that tone; he joked a lot. Manfred then looked up "Eugenics": nothing. "Passion den": still nothing. Frustrated, he slammed the dictionary shut, heaved it aside. Cassell's was Kurt's idea of the best. It was Manfred's, as well. What then was the worst?

Manfred switched the light in the room on and off, looked inside the closet. It was four feet high, perfect for *die Alte*. A desk stood against the wall at the foot of the bed, with a red lamp on it. Alongside the lamp, two pirate bookends rested back to back. The room was six feet wide and ten feet long; Manfred paced it off, measure by measure. On the floor was a shag rug, hideously green. The desk had carvings on it, designs of the sun crudely marked here and there. The rays were hand-grooved into the wood. They looked like swastikas. Manfred checked the drawers in the desk and the little dresser alongside it; they all smelled of wood shavings. While he was putting away his knickers, Florence appeared in the doorway. "A little cocoa?" she asked brightly, holding out a cup and saucer. Manfred stared at her. Her legs were very thin. Her behind was flat. Her buttocks were like two slats. She was smiling. "It's nice to have somebody in this room at last," she said, putting the cocoa on the Indian desk. "I mean, what good's an empty room to anybody?" She paused a moment and looked into his eyes. It was painful to look back. "You have such nice rosy cheeks, you know that? Don't blush, it's a compliment. The girls won't be able to stay away." Then she made a move towards him and tried to take his head between her hands again. Manfred inched backwards. "You don't like to be touched, I'll respect that," she said. "Now stand still, I won't bite. We'll just have to learn how to be patient with each other. Tell me, who do you look like, your mother or your father?"

Words seemed to sizzle in Florence's mouth. Manfred could tell that she could hardly control them. When she spoke, a strange light went on in her eyes. It was some kind of manic energy, like a small scurrying animal's. He could hear her a few minutes later, talking to Adele downstairs again, asking Max a question. He could hear every charged word. Adele had the same light in her eyes, but she didn't talk so fast or so loud. You could see at first glance that Adele was both her mother's and her father's daughter. She smiled like her father, her nose and eyes were just like his, but her tongue, like Florence's, was in a race with her brain. Manfred Vogel knew what that was.

He sat down at the desk now and waited. The cocoa began to cool, skin formed on top. The room was empty of real life, it was like waiting inside a vacuum. New objects surrounded him. They gave nothing back. They were still waiting to be born through his touch. He stared at the pirate bookends, Max's choice. Two bronze cutthroats, leaning on their swords, two swashbuckling silent strangers. What he really should do, he knew, was to write a letter to Frankfurt. Erwin Marks had already written two while they were still crossing the Atlantic. Even Hugo Bondi had been at work on the *Europa*, scrawling brief messages on a couple of ship's postcards for delivery to

Hesse. He should really send a long description of the week's events to Julius and Elsa and his brother. They must be anxious. They were always anxious. Worry came so easily to his mother and father, it was the one thing they didn't have to work for. But his hands were twitching at his sides, his early cheerfulness was gone with his nap, so was the sense of sweet anticipation. There was now a strange aura in the room, not at all like the Mendelssohnstrasse, nothing like Frankfurt, something pervasive and final that would not let go. It seemed to be everywhere, now that he recognized it, spread evenly throughout the skinny little brick rowhouse. Florence Gordon carried it with her, so did Max and Adele, like a personal possession or a gift. It lay in the open cut of their clothes, in their loose, shining hairstyles and their careless uninhibited speech. It came from the noodle-shape of this room and its heavy waxed smell, from the furniture itself, from the peculiar rough-hewn desk, the pirate bookends, and the little blond fellow straining so valiantly on the headboard of the bed. Outside on the street, it persisted. Six of the eight boys, silently tossing the football back and forth under the sycamore trees, wore long pants, and their sweaters had strange primitive patterns on them. None of this had even been suggested by the all-knowing atlas Manfred shared with Kurt at home, so fastidious and so comprehensive in its detailed statistics about America and the rest of the world. It didn't offer so much as a hint, it was worse than Cassell's. Listing all the facts, page after page, it managed to leave everything out. What could a German atlas or a dictionary know? All anybody knew at home was Frankfurt.

Manfred found a little energy at last. He put away the rest of his clothes and arranged his toothbrush, toothpaste, and new razor on the dresser. So far, Manfred only had to shave twice a week. He put the Karl May adventure books between the bookends on the desk, set the Cassell's next to them. Already he had a library. Then he sat down. His eczema itched. He felt for his crooked front tooth. He would have to get rid of his knickers. His knickers were not good enough. They were too German, like the name he had given up.

November 21, a calendar on the desk read. It was the same date in Frankfurt: Saturday night. But it was Sunday morning in Frankfurt, he remembered. Everybody there was fast asleep. His father was snoring, his mother lay in bed with cream on her face. His family lived six hours in the future; his life would always be catching up with theirs.

A radio came on downstairs. "Stop that noise," Florence called, almost beneath his feet. "It's the 'Hit Parade,'" Adele said. "It's not the 'Hit Parade,'" Florence yelled. "I know when the 'Hit Parade' comes on. It's too early for the 'Hit Parade.'" "How about it, Adele?" Max said. The sound switched off. It was silent again.

Clutching his toothbrush, toothpaste, comb, and towel, Manfred walked to the bathroom, five full steps to the right of his bedroom door. He sniffed the air for the strange new aura. It was everywhere. He knew that it would always be everywhere. It floated up from downstairs, from Florence's kitchen, from out each bedroom. Standing uneasily at the bathroom door, he inhaled deeply and made a face. He smelled cheap lilac water, laundered linen, astringent new American soap. Chewing gum and stale cocoa. Gordon smells. Gordon things. The bathroom itself was very modern, done all in white tile with a few black inserts for contrast. That was another part of the strangeness. There was a bathtub with a shower exactly like the one in the hotel in New York. He flushed the toilet experimentally, tested the toilet paper between his thumb and forefinger. A Bromo-seltzer bottle, a tin of aspirin, and a bottle of eau de cologne sat on a shelf. Slowly he went about the business of brushing his teeth, squeezing the white paste onto the bristles in a thin coiling snakeline. The toothpaste was made in Frankfurt and smelled of licorice. Almost everyone in Manfred's family used it. They all smelled faintly of licorice whenever he kissed them. Manfred moved the toothbrush in slow deliberate patterns, scouring away fiercely in the merciless bathroom light. He tasted the licorice, swallowed some foam, working the brush hard. His right wrist began to ache; he even drew a little blood. Then he rinsed his mouth with slow, throaty gargles that he had learned from watching Uncle Benno.

When he finished with his teeth, he went to work washing his hands. The hot water ran on the instant, the soap frothed hydrophobically. He checked his nails, paid attention to his cuticles, everything became pink again. Then, while his hands were still soaking, he began to scrub at his face. He was very rough. Cinders washed off, train grime, a bit of old chocolate. He did his ears, each rococo groove carefully scooped out, and just as carefully probed the pores of his nose, looking sternly for blackheads. Finally, stepping back from the basin, he stared at himself in the mirror. His forehead was shiny, so was his nose. Otherwise, he saw that he had a high, ruddy color (not like

Adele Gordon's flat American pallor), with appetizing rosy cheeks that no one would be able to resist. Without thinking then, he picked up the tube of licorice toothpaste and dropped it in the wastebasket. There was already a pile of used Kleenex at the bottom, covered with lipstick marks. Manfred stood still, gazing at himself. He saw half-a-stranger staring back, a pleasant German face preparing to shed itself. He began to comb his hair, feverishly struggling against his wicked cowlicks. Nothing worked. Then he decided to change the part in his hair from the right to the left side, realign it cleanly, but he had trouble making it straight, weeds seemed to be growing everywhere. When he finally had it the way he wanted, it didn't look bad. It angled his face a little more sharply, made it thinner. He liked it. It was both strange and familiar at the same time. He now had a new part and a new name. He was scrubbed clean, too; he was immaculate, exactly the way Elsa Vogel always wanted him to be. Everything was washed away for the moment, dirt, train grit, the rubble of Frankfurt memory. He was like a newborn cipher. Softly now, he tried to make a loon sound, like Adele Gordon and her friend down the alley, wailing briefly and unsuccessfully, then put his fingers to his lips and blew an extravagant kiss at himself in the mirror. At least this deathless image of his own vanity made him smile again.

"Supper!" a voice shrieked from downstairs. "Come and get it!"

Later that night—after dozing off over his baked apple at supper—Manfred finally made it upstairs into the yellow bed and was dead asleep within a couple of minutes. He stayed that way, snoring lightly in the skinny room, until breakfast the next morning. In all, he slept for eleven unbroken hours.

Chapter Two

When he awoke, certain patterns were revealed. First, Manfred discovered that it took the entire morning to read the Sunday newspaper in America, beginning with the news section and the weather report, then moving on to sports, entertainment, rotogravure, society, and a few odds and ends, including the classifieds, which were tagged onto sports. The sections were passed from hand to hand in the living room, Max to Adele, Adele to Manfred, and then back to Max. "Don't crease it like that when you're reading it," Max said to his daughter, impatiently. "Have some consideration for the next person." Sitting on the sofa at the other end of the room, Adele looked put upon and hid herself behind the paper. "Did you say something?" Max asked suspiciously. "Oh, daddy, cut it out, please, it's Sunday morning." Then she hid herself again, while Max cleared his throat noisily. "Little-orfulannie," he said, reading on. As for Florence and the Sunday papers, her habit, she explained to Manfred at breakfast, was to wait for Sunday evening and have the whole mess to herself while the rest of the family listened to the radio. Sunday night was radio night on Fairfax Road. Florence could read the paper and listen to the radio at the same time. It was easy, she claimed, if you just had the knack.

Among the odds and ends in the paper were the funnies. That was what Adele called them, when Manfred asked. On Sunday, she said, they were in color and each funny had at least half a page to itself. After putting sports aside and then rotogravure, Manfred took a full hour to get through the funnies, hacking his way through the slang as if it were a kind of prickly underbrush, going slowly all the way.

"But why are they called funnies?" he asked Adele. He was kneeling over the paper on the floor, stretched out on a blue rug.

"Because they're jokes," Adele said. The society section lay crumpled in her lap.

"They're not jokes," Manfred said.

"You know what I mean," Adele answered.

"No," Manfred said. "There are no jokes in 'Buck Rogers.'"

"Oh, well, 'Buck Rogers.'"

"They're cartoons," Max said, from his easy chair, carefully working his way backwards through the news section. "They're caricatures."

Manfred looked up. "It's not funny," he said.

"They're not all supposed to be funny," Adele explained. "They're adventure, that kind of thing."

"A woman hits her husband on the head with a *rollholz*."

"A *what?*"

"A *rollholz*."

"Let me see that." Manfred showed her the strip. "That's a rolling pin," Adele said. "That's a joke."

"But next to it look how nice it is," Manfred said, pointing to "Skippy," who was sitting in the public library this morning, trying to learn *Treasure Island*, his favorite book, by heart; even nicer was the fact that he had a headful of stubborn cowlicks popping up all over the place.

"It's nice I guess," Adele said. "But you're too old to be reading that stuff. That stuff is for kids."

Manfred stiffened a bit. He would have to be a little careful. Adele could pinch. He had learned that in less than twenty-four hours. She was one of those who spoke without thinking. Consequences didn't matter to those who spoke without thinking. Consequences were for other people. Adele was something like the funnies, maybe, sharp and soft at the same time, dreamy and real, funny and serious; she blew kisses to the moon and made wisecracks; there was probably a *rollholz* always hidden somewhere on her, as a joke. Manfred turned away then, almost imperceptibly shifting his position on the rug towards Max. There lay safety. Max was kind and generous, sunk deep in his great velvet wing chair, even though he sounded a little cross this morning. Manfred knew he could depend on him, crazy driving and all. Hidden there behind his newspaper, he made purring sounds as he scoured each page; a sort of profound hum of approval or gentle disapproval came from him as he finished one story and went on to the next. Manfred edged closer across the worn blue nap; away from Adele Gordon. He could now make out the headlines on the front page Max was holding. He read them out of the corner of his eye and discarded them at the same moment. He had seen them before, he had been reading them all his life. Front pages were alike everywhere, hysteria and black pain, the world in frantic disorder, threats, omens, cosmic warnings. They made his heart turn over. Almost at Max's feet at last, he began to hum under his breath, beneath Max's own hum, and, kneeling close, eyes down, started hopefully on the "Gumps." The Gumps were having a picnic; ants were attacking the food. Manfred smiled to himself.

"Hmm," Max said, examining an inside page of the news section, right over Manfred's head.

"Find something?" Adele asked.

"Woman falls on man in jump from hotel," Max read out loud.

"There was one like that last week," Adele said.

A few minutes passed. "Listen to this," Max said. "Food within reach, woman starves to death."

Manfred looked up from the "Gumps." The ants had won, the Gumps were routed. In the news section, a woman starved to death. Headlines drifted in and out of his vision. He turned around. Across the room, the morning sun caught Adele's coppery hair. Her skin, framed there, looked very white. She had cleaned her brown-and-white shoes.

"Man found drowned in Maryland creek," Max read, enunciating clearly.

"Standard stuff," Adele stated. She was turning a page in the society section, first wetting her forefinger.

"Poisonous snake kills owner, eight," Max continued. "How about that?"

"More like it," Adele judged. "Everybody loves eight-year-olds and hates snakes. That's a winner . . ." She turned another page. She wore ankle-high socks made of wool. The rest of her legs was bare. "What is it with this paper?" she asked. "There's not a single Jew in the wedding announcements. It's all Smiths, Sydenhams, and Igleharts. From Gibson Island. Where is Gibson Island, anyway? Who cares about Smiths, Sydenhams, and Igleharts?" She pronounced it Igglehart.

They were silent again. Adele turned a page, so did Max. Florence was still upstairs, putting on makeup. They were all going to visit Aunt Clara and her husband, Lester, so that they could check out Manfred. Manfred sat on his haunches on the floor, his heels dug into his buttocks. He was struggling with Tailspin Tommy. "This was supposed to run next Sunday, not today," Max suddenly called out. "How stupid can you get?" He held the paper out. "I'm not going to pay for a wrong insertion," Max said. "They've tried to pull that before." Manfred got up, Adele came over and stood behind her father's chair. "Pretty classy," she said.

"What is it?" Manfred asked.

"Read," Max said, pointing to the upper-left-hand corner of the page.

"'Save With Schick!'" Manfred read. "'No Lather! No Brush! No Blades! $15.00. 50¢ a week. No interest on carrying charges. Mail orders, too! M. Gordon & Son, Jewelry.' It is you?" Manfred asked.

"It is *us*," Max answered.

"Who is the son?"

"A figment," Max said. "Designed for commercial respectability. It gives it an institutional feel, something solid. People like that. It's a test to see how many returns we get. Want to place a couple of bets? You'll help me check out the orders downtown, Manfred."

"Oh, you'll love it down there in the store," Adele put in. "Two by four."

"I don't see you complaining when you need a bracelet or something," Max said. "The ad looks OK, considering. Shall we make a pool, the three of us, and place some bets?"

Florence walked in then. "We're supposed to be at Clara's in an hour," she said, checking her watch. "So no last minute rush, please. I can't stand rush. There goes Hilary again." She cupped her hand around her ear, as if she was having trouble hearing the banging from next door. "If they'd only tune the piano. Who was reading the paper on the floor? Pick it up and make it neat for the next person." While Manfred jumped to put the funnies together, Florence went around the room, plumping up pillows. She also began to move each piece of furniture, but only a quarter of an inch. "Did you write home yet?" she said to Manfred. He shrugged. "What does that mean? Go upstairs and write your mother and father a letter. I bought some airmail stamps for you, you can drop it in the box on the way. Go upstairs now. Pay attention. You have one hour. And as for you, Miss Huchenlepper, move over and let me get at the sofa. Look at the mess. And for God's sake take down those balloons," she ordered, glancing out the window and laughing. "All the air's out of them. You know what people'll think they are. Max, did you shave?" Max put his hand to his jaw. "You call that a shave? Where are the breakfast dishes? Adele, are the dishes still sitting in the sink? *Adele* . . ."

Upstairs, Manfred sat at his Indian desk, cleaning his fingernails with a ruler. A sheet of stationery lay in front of him, his pen was beside it. "Dear Momma and Poppa," he had written. At the top: Sunday, November 22. He put the ruler down, picked up the pen, chewed on it. "It is nice here." He could hear the vacuum cleaner roaring downstairs. "Mr. Gordon is very nice, especially. Sometimes he is very amusing. The lady of the house is a little noisy"—downstairs now, talking over the vacuum, she sounded like Minerva Gump—"but all in all very interested in my welfare. Of course, it is not Frankfurt, you understand that. Here the streets are paved with gold, and there are Indians everywhere. But I refrain from making judgments. It is too early. Everything is too strange. I will wait three days before I make a judgment. I have never seen an Indian before or walked on gold. On my first night in New York, Mr. Gordon took me to a restaurant with dancing. Tell Tante Ernestine that Ginger Rogers was there. Adele Gordon has big teeth and she likes to smile. The house is small and neat, very compact and comfortable. So is my room. Let me tell you about the sea crossing . . ."

A few paragraphs on: "If you ask me, English is a language full of small sounds. Pa, pa, pa, pa, pa, like a machine gun. I get a headache from speaking it all day. I think that is probably natural. I am getting a headache now from writing in German. My head is like a revolving door. English in, German out, at tremendous speed. There are collisions. Kurt will understand. It is not like school. It is real life. Well, I see in the papers this morning there is trouble again. In Baltimore, a child was poisoned by his own snake. Imagine. Is it true that forty professors in Hesse asked for a ban on the Olympics? That is what the paper says here. The paper also says there was a pogrom in Klwow, ten dead, and Kurt, according to the Nazis, is now decreed a 'mongrel first-degree.' I always knew he was a dog. Speaking of jokes, I read one in the newspaper this morning. A woman named Dorothy Dix writes that vice is the poorest paid of all occupations followed by women because it lacks permanency and seldom has a future. If you don't understand, ask Kurt to explain. Well, I must say it. I love you with all my heart. Finally, it is easy for me to say, four thousand miles away. I have never said it before. It is strange, strange. It is like I am in a trance. It is like I am half-asleep, walking through my own dream. Do you understand? I will wake up tomorrow, I am sure. Tomorrow the dream will be over. No more trance. No more sleepwalking. You must come here. You will like it. All of you. It is not Frankfurt but the streets are paved with gold and there are Indians everywhere. I am *really* all right..."

Just before they left the house, Florence gave Manfred a sweeping look, head to toe. "One of your flies is unbuttoned," she said. "Florence," Max objected. "Then you tell him," she said. "You always turn me into the bad guy. He's not made out of eggshell. You're not made out of eggshell, are you?" she asked Manfred, who had turned his back to button his pants. "Take it easy," Max said to the whole room. "And quit snickering," he added to Adele. "Don't mind them," he said, when Manfred turned around again. "They always get this way when we go to my sister's. They get crazy. Now, where are the car keys? I left them right here on the radiator yesterday. Come on, who has the car keys? Nobody move..."

Elapsed time: Frankfurt–Bremen:21 hours; Bremen–New York:122 hours; New York–Baltimore:44 hours.

A-mer-i-kha.

Chapter Three

Driving out to the Ottingers, Manfred asked for coed and got it, without a fuss. There wasn't even a discussion, although he had prepared himself for one. Adele had skillfully laid the groundwork with her parents, and the decision was quickly made as they drove up Garrison Boulevard. Of course, Max said, and Florence agreed. Manfred would spend a couple of uncommitted months at Forest Park High, until the term ended in February, when he would be ready to join the real sophomores once the school had learned what his aptitudes were. That was the plan. Clara liked the idea of coed, too. It made sense, it had a rationale. Also, it was more natural, she told Manfred when they all finally reached the Ottinger house, it reflected the world as it was. Her theory was that Manfred should learn about the realities of American life, the sooner the better. Coed was one way. Working downtown at M. Gordon & Son was another. A third, she pointed out with a complacent little laugh, was to get rid of his new European beret, which a black maid had just hung up in the front-hall closet. Nobody Clara knew in America, certainly nobody in Baltimore, Maryland, wore a beret. Just put it in a bottom drawer and forget about it, was Clara's easy advice.

Clara and her husband, Lester, lived way out in Mount Washington in a sprawling white house made of clapboard. The house was set on a half-acre of sweeping lawn, facing Pimlico Racetrack a couple of hundred yards away, across Rogers Avenue. A huge porch, hung with honeysuckle vines, enclosed the house on two sides. Giant firs stood out front. Fat hydrangea bushes marked a path, a rose trellis ran up the garage wall. The vines, the bushes, the trellis were all bare now in early winter. The Ottinger house was not at all like Fairfax Road, Manfred discovered as soon as they walked through the front door and gave up their coats and hats to the maid. "Stand up straight," Florence whispered to Adele, who instantly pulled in her stomach with a great sucking sound and stood there in the hall for a second or two, rigid, her cheeks puffed with air. "Act your age," Florence ordered, under her breath. "You always start carrying on like this." Clara's and Lester's house contained both a living room and a library and, facing west, a sunroom filled with rattan furniture from Singapore. Dark paintings of American pastoral scenes—cows, meadows, sunsets, and Hudson River palisades—hung on the walls of the living room, each one specially lighted from above; yards of Victorian literature, all of Dickens and Trollope included, lined the shelves of the library; and an ebony Steinway six feet long, used for occasional at-home chamber music recitals, stood alone, somewhere on the first floor, with an adjustable leather stool sitting in front of it. There was also, to Manfred's amazement, a kitchen in the back of the house that was tiled in blinding white from floor to ceiling. Each tile was like a patch of fresh snow. "That's Shorty there," Lester said to Manfred as they stood in the kitchen door together a few minutes after they had met. He pointed at a small black man who was putting glasses away. The black man had a hump on his back. "Shorty outranks me in this house. What he says goes. Right, Shorty?" The black man, blowing one of the glasses dry, laughed. "He thinks it's funny," Lester said, laughing himself. "Because he knows it's true."

Manfred learned from Uncle Lester then that his new cousin, Babette Ottinger, was away at Bryn Mawr, where she was a freshman. Babette's older brother, Stewart, was at the University of Pennsylvania, studying to be a doctor. Bryn Mawr was just a hop-skip-and-jump, so to speak, from the U. of P. This was very convenient, Lester explained at dinner to Manfred, who was seated at his right, as the guest of honor. It was also comforting. The Ottinger children could see each other from time to time in Philly. They could have a meal together downtown at Buchbinder's. They usually did on Friday evenings. At their father's expense, of course, Clara put in from the foot of the table, with good cheer. She interrupted Lester all through dinner, with little footnotes of enlightenment, and interrupted her brother, Max, too, who seemed to be used to it. Sitting straight in her chair, Clara was taller than both Lester and Max, just enough to make a difference. Her hips were slimmer, her shoulders broader. Her hair was clipped so short in the back that Manfred could see the bristlemarks on her neck; from there it rose in precisely shaped waves; otherwise it was like a man's. She had a man's authority at the table, as well, brisk and sure of itself. Manfred thought that Max Gordon was a little afraid of his sister.

"I assume you've heard that I'm originally from Hamburg," Lester said to Manfred later, showing him around upstairs. "Not literally, maybe, but I am a kind of a refugee myself, so to speak, what you might call a remnant of the '48 revolution." Manfred had to stop a moment to figure this out, it was so unexpected; the facts of German history began to rattle dimly around his brain. Forty-eight, he thought, trying to focus. Without expression, he glanced at Lester, who stood there with his thumbs hooked into his vest, gazing at him solicitously. Lester's eyes were pale blue. They were also mild and questioning. Ever patient, he seemed to count to three after each remark before going on with what he was saying. "As I said, not me literally," he continued now, in a softly didactic voice. "I'm speaking of my grandparents." He turned to walk down the long hall, Manfred following him. "My family has been Deutsche for generations. Did Max tell you that? There are a lot of Deutsche families around here from the '48 revolution. They saw another future, back there in Munich or Hamburg or wherever. They ran for their lives. They saved themselves by believing in their own good fortune, by believing in themselves, and most of them became successes because of it. Not bad, in my opinion. Over the years, of course, they've turned into social conservatives and, in some cases, social snobs or worse. Money made that happen, new money and the fear of being taken for Jews. Jews from the east, that is. Jews with the wrong accent. I broke the social pedigree when I married your Aunt Clara, whose papa was a Polack, if not worse." Lester chuckled to himself. "I went against the Deutsche grain and tainted the Ottinger blood forever. Imagine, I had a father-in-law who could speak Yiddish. Are you listening to me?" Lester asked, looking over his shoulder. "Does all this interest you? There are German Jews in this town who still haven't forgiven me. It's well known that I have always been attracted to a certain kind of social martyrdom, a few thorns and nettles on occasion, but no blood, and naturally the world has been only too happy to give me what I wanted.

"I'm just a little irregular," he went on, after another pause. "It doesn't take much to be a little queer in this town, where regularity is prized above most things. Bowel movements—you know what I mean?—marriages, genes, money, and adultery. Everything has to be in order. Everything has to fit. You can break the rules from time to time, but you can only break them according to the rules. Anyway, we're rich, your aunt and I, not to put too fine an edge on it, so we get to call our own shots. Mustn't forget that, it's crucial. We have our cake, and we don't hide it beneath a bushel, so to speak. As though you couldn't tell," Lester added.

Manfred stared blandly at his host. He didn't answer. He had no answer. He was struggling desperately to keep up with Lester's monologue, trying to impale the impressive vocabulary before it vanished and hold onto the figures of speech, which went by at such speed, despite Lester's pauses. It took intense concentration, everything was so fast. He'd have to work out the meanings later, try to reconstruct the flow of words. They came back downstairs, Lester leading the way. "You know, I've never had to work for my living like most other people," he said. "That's the essential fact of my life. It's dead center, so to speak. So I've been alone a lot. I mean I've had to fill the time that would otherwise go for work, kill it, in some instances. People don't think of that when they think about independent means. They only think of freedom, not the boring part. And it hasn't been made easier by the fact that I'm not so good at finding cronies. Nothing and nobody's ever good enough for me. There's always a hair in the soup. I can't help it. It's my way. Maybe having money did it. It usually does. The stuff came from Ottinger Luggage. Made in Baltimore, Maryland, carried all over the world. Class merchandise. But the stuff won't last. Not by itself it won't. I mean my kids will have to work to support themselves, Babs and Stewie both. There won't be enough money for them to live independently like their old man. I haven't lifted a finger on their behalf in that line, beyond barest duty, and the depression hasn't helped any. So they'll go out and make their own livings and they'll forgive me for it. No, more, in fact, they'll learn to thank me for it, they'll be grateful, it's a privilege to be able to earn your own living, although I'm probably not the one to say so. The important thing is my life has been blessed by a few people I love. Some of them even love me back. That's luck. That's privilege. I hope you'll be one of them. Don't let anybody tell you I'm a dilettante. It's the family line. I am not a dilettante. I'm a connoisseur. I know about things, about people. I know what goes on between them. I have intuition and I see clearly. I've been watching a long time. I hope you'll come to appreciate that. It's worth appreciating. It really is. You get the whole picture now? Too high-flown, maybe? I'm sometimes prone to highfalutin talk."

He laughed again, modestly, as though all these amusing ideas had just occurred to him, right after they all got up from the dinner table. This time, Manfred laughed with him, out of an irresistible sense of duty. He could hear his own laugh come back at him in the Ottinger hallway, sharp and edgy, overlapping Lester's. He didn't want Lester to dismiss him, he wanted to impress him with his wit and his sophistication, but his new uncle made him nervous. All Americans made him nervous. Lester made him nervous, even as he made him laugh. Manfred patiently waited for his cues. In front of him on the wall hung a small etching of a man in profile wearing a stovepipe hat, signed Manet. Manfred stared at it, still waiting. The monologue was over. His uncle, his virginal face clean-shaven and unemphatic, was looking at him from under half-closed lids. Manfred glanced again at the etching, glanced at Lester. He continued to wait for a cue. He wanted to be accepted in that house. He wanted to tell Lester—in the right accents—that he knew who Manet was, but no words surfaced; only a pained shyness.

Hair-in-the-soup, he thought.

On the ride home later in the afternoon, Adele and Manfred were sprawled side by side in the back of the car. "He talk you deaf, dumb, and blind?" she asked, poking him in the ribs.

"Who?"

"Uncle Lester. He never shuts up." Adele herself had only spoken about a dozen words at the Ottingers, maybe less.

"I have an uncle like him in Frankfurt," Manfred said, looking out the window.

"Lester's rich."

"So is my uncle."

"Rich people think they can say anything they want. It's a class trait."

"You don't do so bad yourself," Florence said from the front seat.

"I don't talk people deaf, dumb, and blind."

"Dumb is enough," Florence said.

Alongside her, Max stopped at a red light without stalling. He lit a cigarette and turned to his wife. A moment passed. They all heard him breathe in deeply. "I'm thinking of expanding," he suddenly exploded in a loud voice. He touched the hair at the side of his right ear, blew smoke at the windshield. He looked surprised at what he had said. His words hung in the air ambiguously. "It's the right time," he added, as though he had been challenged. The light changed, they moved on. "Hear what I said?" he called, looking at Florence and the road at the same time. Boarman Avenue flashed by; Belle, Sequoia. "Florence?"

"I heard."

"Then say something."

"What are you thinking of expanding?" Florence said, in an innocent voice, finally turning to him.

"You know, sometimes I don't believe you. I really don't."

"Watch where you're going."

"That's all you have to say?"

"What do you want me to say?"

"Here we go," Adele whispered, poking Manfred in the ribs again.

"It's the right time, Florence, I know it." Max hit the steering wheel with his fist. "These opportunities don't come twice."

She turned away again. "This is no time for expansion. If you ask me. Are you asking me? You still have the fixture loan to pay off."

"Oh, the fixture loan." Max hit the steering wheel again.

"Let some of that smoke out the window," Florence said to her husband. She coughed into her hand.

"You don't like the idea," Max said. "I told Lester it would be like that. I told him you'd be against it."

"Well, if you have all the answers, why do you have to ask the questions again?"

"I knew it. I should leave well enough alone."

"You always get restless when we go out there. It never fails."

"I have a surplus this year," Max said.

"You still have the fixture loan to pay off."

"I have more than I need."

"So bank it."

"I have ideas for the store. You know that, that's not news. I'm not restless, I'm ambitious, what you always used to say you admired in a man."

"Your sister has that effect on you. It's always the same. But Lester's rich, what do they care? Easy come, easy go."

"This has nothing to do with Lester. I don't have to be rich, that's not what I'm talking about. I have a few dreams, that's all. Better, I have a few plans. Real ones. My feet are on the ground."

Florence was silent.

"It isn't the money," Max went on. "Or only the money . . . Florence?"

"We can talk about it another time."

Adele poked Manfred again. He snuggled down into the seat, Adele snuggling alongside him, shoulders and thighs touching.

"You'll excuse me, I don't think this is the occasion," Florence said.

"Where are the songs of springtime, Florence?" Max asked in a desperately cajoling voice. "Oh, God," she groaned. "Not that. Save me this once." The car began to jerk along the trolley tracks as Max disengaged the clutch. Adele and Manfred rocked against each other. Florence leaned forward and turned on the radio. The New York Philharmonic was playing Beethoven. Massive chords thudded in the car. Manfred beat time on his knee with his fist. His new beret lay in his lap. Adele was chewing on a nail, her head resting on his shoulder, as though it belonged there. "Damn," she said, as a piece of skin came off. But her head stayed where it was. He thought he could feel the feathery sheen of her hair, the alert coppery tone, brushing his chin in the dark. He hardly dared move. Max slowed down for another intersection, started to speak. "Double-park across from Silber's," Florence cut in over the music. "I need some bread for Adele's lunch tomorrow. Adele, hop out and pick up a loaf. Adele? What are you two doing back there, anyway?"

"We're making friends. That's what you said I should do, isn't it?"

"Friends," Florence said. "Well, quit sitting on your spine like that, I've told you before, you'll get curvature. You too, Manfred."

"Sit up straight," Max said, fooling with the rearview mirror. "Both of you. Do as your mother says."

"Make sure the bread's fresh now. Don't let them palm yesterday's stuff off on you. You have some money? Here,"—thrusting out her hand—"take a quarter."

Chapter Four

They put him at center his first time out—it seemed neutral enough—but when he hiked the ball, growing dizzy looking at the world upside down, it either flew onto somebody's lawn or skittered up the street, along the asphalt. Everyone groaned, but with restraint. Their parents had told them to be polite, to welcome Manfred, to try. So when the ball landed in the crook of a sycamore branch and got stuck there, they made him an end. Joel Friedberg blocked him with his elbows, shoving him aside without much effort, just spinning him around easily, and even when Manfred did get loose—it happened twice—he had trouble running and catching the ball at the same time. The ball was too long, too slippery, too foreign; it bounced out of his arms like a swollen egg; and he was afraid of tripping over Joel Friedberg or somebody else and banging his head against a car fender. Still, he kept trying. He, too, was polite, he wanted to be welcomed, even though after a week all the kids still looked alike to him, with one or two exceptions.

Finally they put him in the backfield with Adele. He liked it there. He didn't have to face Joel Friedberg's hot breath in the backfield, or his sharp bruising elbows. He felt that he could see everything as it happened. And Adele's bossy presence alongside him helped to give him confidence. While the play went on, he mostly jumped around in his square-tipped German shoes, outside the action, making strange throaty noises as Jojo McAllister raced down the street with the ball clutched under his armpit or Billy Brent, his blond cornflower curls damp with sweat, got ready to pass. Billy Brent had a habit of feinting with the ball, running backwards on his toes with a teasing half-smile on his face that kept everybody guessing. Billy was one of the kids who had a face of his own.

A couple of times, they let Manfred do the kicking. "Jee-sus," Billy Brent kept saying, as the ball rolled limply into the gutter a few feet from where Manfred stood. Then he looked down at Manfred's feet. "Where did you get those shoes, anyway?" he asked, while Manfred blushed. Occasionally they let him run with the ball. "Holy shee-it," Billy Brent shouted, each time he fumbled. "You got butter on your fingers or something?" "Oh, come on," Adele said. "Take it easy. You're not exactly Buzz Borries yourself." "I don't claim to be Buzz Borries," Billy yelled. "I just want to play a little ball." A moment later he sidled over to Manfred and patted him mournfully on the shoulder. "OK, kid," he said, his teeth gritted for politeness, "let's get moving." Obliging him, Manfred began to jump around again. "For Christ's sake, Moon-fried," Adele shouted. "Quit acting like a dervish. Stand still and pay attention."

Throughout the game, Jojo McAllister stood on his head at arbitrary intervals, as if he had to drain his body of poisonous substances every ten minutes or so. His face turned purple. His pants slipped over his ankles. Spittle ran down his chin, pennies fell out of his pockets, followed by his housekey and a couple of aggies. Jojo sometimes even stood on his head while Adele or Joel Friedberg or Billy himself were heading for a goal. "Fucking-A," Billy commented bitterly at one point, standing with his hands on his hips, against a car. Jojo's short legs were wavering in the air. "You're headed for a freak show, for sure," Billy said.

When Adele ran with the ball, Billy Brent always seemed to be in her way, somehow, holding out his long-fingered hands to grab her around the waist or grapple her to him sharply by the shoulders, shouting "Gotcha!" or even—once—to swipe a feel of her breast. When this happened, Billy's eyes looked a little wild, as though he had surprised even himself, but everyone else, including Manfred, turned away, pretending to be doing something else. A few minutes later, ready for him after a particularly wrenching thrust, Adele gave Billy a frog right above the elbow that stayed black and blue for forty-eight hours.

Manfred got to call the signals once during the first week. In all the tumult, everybody talking one on top of the other, he began to think in German and English at the same time. Their side huddled together in the middle of the street, arms around each other's shoulders. Joel Friedberg was breathing hard next to Manfred. Jojo McAllister spat into the circle. "Somebody decide," Adele ordered, "besides me for once." "I got it," Jojo said. He spat again, looked serious. "Vogel," he said. "You run off to the left and I'll toss it. Just count one-two-hike and take off." Links, recht, Manfred thought.

One, two. He fumbled again. The play failed. George Reese, blowing his nose, said "Shit." Jojo stood on his head against a tree and farted at the sky. A delivery truck from Hutzler's, creeping down the street, sent them all off to the curb, where they stood in a ragged line, looking affronted until the truck passed.

The next afternoon Billy gave Manfred ten minutes of individual coaching before the teams gathered. He wore a maroon sweater with geometric Indian designs running up and down the front; a day-old beard stubbled his chin; his mouth, which seemed painted on, was like a closed rosebud; he was eighteen and an even six feet. Manfred stood there stiff with fear and hope while Billy showed him how to hold a football for passing. Like this, he said, seizing it in his huge hands, so that it'll spiral. Manfred watched him with watery eyes. This way, he told Manfred, it will fly like a bullet. When Manfred tossed the ball, it always wobbled slowly towards the wrong player and then fell short. "No," Billy Brent said, pulling at his crotch in anguish. "Grab it here. Jee-sus, just get hold of it, fingertips over the laces, your hand's big enough for that." How-to-throw was followed by how-to-catch. By now, Manfred was perspiring and the sweat was making his eczema itch. "You just bring the ball in with both hands," Billy said, "close to the body, like this, nothing to it. Remember," Billy added, wrapping Manfred's hand inside his own as part of the demonstration, "it's not a round ball. It's, uh, obloid. It's made so that it'll fit neatly against your side. You see? Right in here, curve against curve." He brought Manfred's enclosed hand down against his hipbone, kept it there a moment before dropping it. "You cold or something?" Billy asked. "You're shivering. Come on, get out there, I'll pass you a couple."

The players met after school each day. Everybody got smeared with sweat and dirt except Teddy Lewin, who played halfheartedly because he was afraid of damaging his piano fingers. Soon they all began to think they were going blind as the light dimmed a little bit earlier each afternoon. Where the hell's the ball, Adele kept yelling, peering nearsightedly down the street; by then, it was only four-thirty. The third time out Manfred made sure to tear his knickers, not too much but enough, and felt better when he saw the hole at his knee; he shared that, at least, with almost everybody. In most of the games Adele called the signals, snapping them out in a hoarse, barking voice from her place behind the line. Her authority during the games, which no one but Billy Brent ever questioned, made her stand straight and hold herself at attention. From the curb, meanwhile, where she retrieved out-of-bounds balls if she was in a generous mood, Laura Piscitelli from down the street called out sarcasms to both teams and beat her arms against her body to keep warm; she was almost always there, sooner or later in the day, sometimes with little Kitty Dean in tow.

They made a lot of noise out in the street, screaming and arguing. Neighbors complained all the time, adding to the clamor, and once the cops showed up, looking bored. Sometimes Billy passed around frayed little books that they all tried to keep from Manfred. Oh, no, you don't, Laura Piscitelli kept saying, hiding the book behind her back, nothing doing, it's bad for your digestion. There was always a little blood on somebody's knuckles or knee, scabs that kept opening up, and bitter ongoing fighting over yardage claims. Adele almost tore a nail off with a bad catch, Billy developed a charley horse, and Laura Piscitelli kept disappearing around the corner with Joel Friedberg or Roger Baer as soon as dusk set in. But most of them never missed the game, never gave up an afternoon; some passing sense of pride in their team held them; for a brief chilly season, as the game went on, they all worshipped the filthy misshapen pigskin as though it contained the essence of their very souls. Even, dimly, Manfred Vogel.

Excerpts from a letter, begun in English, from Manfred Vogel to his brother in Frankfurt:

". . . Already, I feel like an American, red, white, and blue. They call me Man-fred Vogel, flat *A*. I don't even know where Germany is. Mr. Gordon took me to the bank and now I have my own savings account with fifty dollars. It collects 2.5 percent interest every year. I also have my first American shoes, brown and white, called *saddle*, very good for sports. Yesterday, I had my first interview at the agency on Monument Street. On Monument Street, there is a memorial to George Washington a hundred feet in the air. His sword sticks out in front of him like a hard-on and that is why George Washington is known as the father of his country. I don't like such interviews with strangers. Is Manfred Vogel happy in America? Does MV like his new home? Does MV get along with the Gordons, with Mr. Gordon and the lady of the house and the girl, Adele? I sound like a spy. I sound like *die Alte*. They can't even remember my name, the woman keeps looking at the file on her desk to make sure who I am. All the time, Mr. Gordon sits outside on a long wooden bench, waiting for me, and later, after we leave, he doesn't ask a single question."

Then Manfred suddenly tired of English and switched to German without warning.

"Mr. Gordon has a jewelry store, as you know, so small, like a fish tank, long and thin. It is downtown, on a busy street. There is a little counter inside, a narrow space for customers in front of it, a small back room, and a safe like an Abwehr tank. Mr. Gordon lets me use his special eyeglass to look at the jewels. In every jewel, a light swims. I learned how to work the cash register. Mr. Gordon showed me a pearl necklace, on order for a rich person uptown, and a couple of diamonds. Chips, really. Then I checked the mail and the orders. I worked there for two hours in the back room. He wanted to pay me fifty cents for each hour, but I refused. How can I take his money? He says where will I get spending money for myself? It is a sensible question.

Tomorrow is Thanksgiving here and soon I will start school. Thanksgiving is an Indian holiday. I have an Indian desk in my room with strange designs on it. In Germany they are called swastikas. (The truth.) Shirley Temple came to dinner last night and Mrs. Roosevelt will be here for Thanksgiving tomorrow. Already I am so important. (Ha-ha.)

There will be Tristan, Bohème, and Carmen at the opera in the spring, the newspaper says . . .

Also, I keep a notebook for words and expressions that are not in Cassell's. With all due respect, English all day is not like school. 'Look what the wind blew in!' Also, 'In one ear and out the other!' 'His eyes are bigger than his stomach!' Also 'The cat's got his tongue!'; and from sports, 'two left feet,' 'goody-two-shoes,' 'bust ass,' 'save ass,' and 'grab ass.' It is good to chew on the American vernacular . . .

The girl of the house is in the room next door. In the morning at breakfast she has puffy eyes and does not speak. She keeps the room closed. She believes in privacy for all. There are things . . .

Here it is just ordinary life. This house, where I live, is overheated, outside it is winter, and the lady of the house talks on the telephone for two hours every morning with her friends. What is happening in Hesse? What is going on? Here I see strange new faces every day, soft faces too, not like Frankfurt, and names that are unpronounceable. I try to hold onto Frankfurt, but the city's spires shrink in my dreams. I told you, Germany grows small. The Schillerplatz has already disappeared. It is all disappearing. Strange? I'm sorry, I don't think so. You will forgive me? Momma and Poppa, too?

Please send photo, wearing white silk scarf . . ."

Chapter Five

On Thanksgiving, they all went back to the Ottingers for the afternoon and ate turkey. It was the biggest turkey Manfred had ever seen; immense; fat; stuffed to the size of three footballs and, once out of the oven, almost the same color. The skin looked varnished and crisp, its own juices oozed and simmered through every crack. Adele had put on a cashmere sweater and plaid skirt for the occasion. The skirt had pleats that were as sharp as a knife-edge. She also wore new loafers, a shiny penny inserted in each one. Florence checked her out a half-dozen times before she was willing to acknowledge that she looked all right.

"She looks better than all right," Max said, a little indignantly, as his daughter spun around in front of her mother; and Manfred had to agree. Her straight coppery hair was combed out and tucked under at the bottom; her face was clean, not a bruise or dirt-mark in sight; she had even put on lipstick that was colored fire-red. All morning she had been charging around her bedroom getting ready. When she finally came out, she didn't look as if she had ever called signals from a backfield. "*Please*, for your own sake, stand up straight, you look so pretty when you stand up straight" were Florence's last words before they entered the house in Mount Washington. "Leave the child alone for once," Max said, ringing the doorbell and walking in. Behind them, grey clouds piled up sullenly over Pimlico. A couple of squirrels skittered across the lawn, disappearing under the firs. The temperature was dropping as they stood on the porch. It looked like snow.

This time, Babette and Stewart were home for dinner. Babs, she was called by her family and everyone else. Stewie. "I see you took my advice and got rid of that European *schmatta*," Clara said, as Manfred stood bareheaded in the hallway. Uncle Lester waved to them from the stairs, on his way down. Shorty took their coats. The floors smelled of fresh wax.

"It's in my drawer," Manfred said.

"You don't need a hat anyway," Clara said. "It made you look very peculiar, with that thing sitting on your head like a floppy soupbowl. Don't forget, this is America, bay-rays belong in Paris, France." Then as Babs and Stewie came forward to shake Manfred's hand, Lester herded them all into the living room.

Once they all settled down, it turned out that Stewie had gone to Forest Park High, like Adele. He too believed in coed. "You'll like it there," he told Manfred a few minutes later, sucking on his pipe. "It's got the right spirit. It's democracy in action. It's archetypical, it makes you believe in public education." Stewie was twenty-three and already studying advanced anatomy at medical school. All his marks were *A*'s. He wore a heavy tweed jacket with leather buttons, imported from Leeds, England, and Argyle socks that came from Aberdeen, Scotland. A Phi Beta Kappa key hung from his watch pocket. Now Babs, he said, turning to his sister tentatively, had not gone to Forest Park. Right? he asked, as if he didn't know the answer. Right, Babs acknowledged. Instead, Stewie went on, Clara and Lester had decided to act on their progressive principles and send her to the Park School, where you paid for your education. "Through the nose, too," Uncle Lester commented from his green leather chair. Pulling his cardigan around him, he got up to make the drinks, including a neat bourbon for himself, which he finished off with one swallow.

At the Park School, Babs suddenly picked up, gratuitously, there were only about seventeen kids in each class. She was talking to Manfred. That's what you were really paying for, she said, a little elbowroom, a chance to expand, to find your own place. Manfred nodded intently. Babs had a very insistent voice. At Forest Park, Stewie had been handed a diploma along with three hundred and thirty-three other graduates, half of whom he had never spoken to in his life. At Forest Park, you were just another cog. While Babs explained all this to Manfred, who was making a face over his ginger ale, she smoked a long pink cigarette, stuck into an ebony holder that she waved around a lot, like a baton. Manfred had never seen a pink cigarette before.

"Please do me a favor," Clara ordered, "and watch that thing. There's already ash on the Ispahan."

Park School, Babs continued, reining in the holder, was built on ideas formulated by John Dewey. "You've heard of John Dewey?" she asked Manfred.

"Dewey was a major influence in American education," Max offered, in an aside to Manfred.

"I heard of John Dewey," Manfred announced, in order not to look stupid. Clara shook her large head appreciatively. The bristles on the back of her neck, Manfred saw, where her hair was cut short, were beginning to grow in.

"Then you have some idea," Stewie said, "what an important educational venture Park School is."

"I thought you believed in public," his cousin, Adele, reminded him.

"I do. That doesn't take away from Park."

"Full of snots," Adele muttered.

There was a moment's silence that lingered. Adele had a reputation among the Ottingers as the family assassin; she was known as a verbal killer. A kind of brief mourning period often followed her remarks out in Mt. Washington. Mild sorrow hung in the air now. Finally, Lester got up and walked over to the drinks-table.

"Well," Clara cried out, while Lester turned his back on them and swallowed a second bourbon. "I have my doubts about Park over all. There's such a gap between the ideal and the reality, with that awful gee-whiz rhetoric everybody spouts. It's all just a wee bit self-congratulatory, don't you think—as if they'd all just discovered the wheel. And careless, too. Babs didn't know the months of the year in the right order until she was a senior, and nobody ever noticed."

"I didn't get into Bryn Mawr on my connections alone," Babs said, sounding peevish.

"You certainly didn't," Stewie said loyally.

There were too many opinions in the room, Manfred decided. All the Ottingers did was talk about themselves. No matter what the subject seemed to be, they were still talking about themselves. Hearing his stomach rumble, Manfred gazed at a painting of a Catskill cow hanging on the opposite wall. Snowy moonlight illuminated the cow; a pool of shimmering water lay at its feet; a broken fence lined the horizon. Uncle Lester was having another drink, this one with a lot of soda in it. "Why don't you just save that one,"

Clara said in her husband's direction, without effect. It didn't seem that cousin Babs would ever address a word to Adele or even turn to look at her. Or Florence Gordon either, for that matter. Cousin Babs talked only to Manfred and the great gold Ispahan rug that lay at their feet. Florence herself hardly spoke a word, just sitting there on the high-backed sofa with a clear, expectant look on her face and her hands folded primly in her lap. Every now and then she would give one of her appealing Gloria Swanson smiles, usually at something Clara had said; sitting in Clara's living room, she approved of everything Clara said. Once outside it was another matter. Maybe, Manfred thought, maybe he should tell Babs and Stewie about Herr Levi and his old Frankfurt classroom with the broken skylight, where the wind rushed in all day and German mice camped out at night. There was plenty of individual attention in Herr Levi's classroom; just ask Joachim Adler.

But it was too late. Shorty was wagging his finger at Clara from the doorway. Dinner was ready. They all got to their feet, famished.

Shorty and the black maid served the turkey on a vast silver platter, which they set out on the sideboard. "Don't fool around now," the black woman said, "and let this gawjes thing get cold." "You talking to me?" Shorty asked. "Oh, you," she said, moving past him. "You get the other stuff from the kitchen and make sure the lid's on all the serving bowls. Move, man." Lester then set to work on the turkey, the black woman at his side holding her breath. Whenever Lester got tired of carving, he took a sip of bourbon. "It's time you did this job," he said to Stewie, over his shoulder. "Just say the word, Dad." "I mean without being asked," Lester answered, glowering at the stuffing.

The black woman's name was Rhea. She and Shorty were husband and wife, Clara explained to Manfred, and they lived over the Ottinger garage in their own little apartment. "You the new young thing?" Rhea asked while Manfred was waiting his turn at the sideboard. "You ain't so little. They says you was little. You taller than me. Try that poorray of chessnut over there. You ever have poorray of chess-nut? You ever have turkey? Get in there and lay your hands on that bird, boy. Don't let Mizz Babs cut you out like that."

With the chestnuts, they had sweet potatoes, which Manfred gagged on at the first mouthful; the potatoes had prunes mixed in with them. He then picked at the spinach, made designs with his fork in the chestnuts, and had two servings of turkey. "A-ha, I see you like drumsticks," Stewie said. "Maybe a little too much," Florence commented, eyeing Manfred's plate. "Nothing's too much on Thanksgiving," Max said decisively. "Eat up, child."

"You know what Thanksgiving's all about?" Adele then asked Manfred, breaking her silence after she had finished her turkey. She had a spot of sweet potato on her cashmere sweater.

"It's about turkey and how many ways it can be ruined," Lester said.

"Oh, Daddy," Babs said.

"Well," Adele began. Then she told Manfred a story about Hiawatha, Pocahontas, and some Pilgrims in Massachusetts.

"Come on," Stewie said, tapping the bowl of his pipe on his heel. "Why are you feeding him that stuff? The story's really so simple. And quite beautiful, too."

"Adele's imagination is always running away with her," Florence said.

"At least it got us through the salad," Adele said from across the table, bringing on another mournful silence.

"Some scotch with your pumpkin pie?" Lester asked Max a little later. Shorty handed him another bourbon. "Don't be a spoilsport now," Lester said when Max declined. "It's a day for excess. Come on, join me." "A-hem," Clara said, while Max shook his head. "You'll all turn me into a solitary," Lester said, drinking up. Shorty began to carry empty dishes out to the kitchen, a bottle of Wild Turkey tucked under his arm. Forks rattled against dessert plates. Somebody passed a carafe of water around. There was the smell of whiskey in the dining room, chestnuts, burned pumpkin. Then Lester got ready to close the dinner—prematurely—with a prayer. One of Lester's eyes was now in soft focus and wandering out of line as he began to deliver the dishevelled words of gratitude. It was over in a minute. Ay-men, Lester concluded. Oh, Daddy, Babs said.

Later, they listened to operatic recordings in the sunroom. Outside, the grey afternoon light vanished with astonishing speed over the Pimlico grandstand. The clouds broke up, a brilliant mauve streak curved through the western sky. While the voice of Caruso rang out through the house, Max and Lester played a hand or two of rummy for a penny a point.

"Cielo è mar . . ." Caruso's voice Manfred certainly recognized. He had heard it in Frankfurt, at Oskar and Lotte's duplex flat, where they let him work the Victrola himself. In the sunroom Manfred sat behind Max, watching him play out each hand. He wanted him to win, he wanted him to beat Lester. Max's white hair covered his head like a spent avalanche. His cards fanned out, Max snapped one in the air and thrust it on the table. Lester hesitated, looking wounded. Manfred leaned forward and rested his chin on Max's shoulder; for luck, he told himself. Hmm, he said, counting up the points. Max was ahead. His jacket was prickly.

"Who's your favorite composer?" Lester wanted to know, examining his hand. "Mine?" Manfred sat up at attention. "Yes, yours." "You mean opera?" Manfred asked. Lester nodded and played a card. "Puccini." "I might have guessed," Lester said. He hummed a tune and looked at Manfred from under his lids. "Name the aria," he ordered. "*Vissi* . . ." "Never mind," Lester interrupted, laughing, and went on with his game.

Clara sat in a deep chair nearby, crocheting a dress for herself. Florence was feeding her gossip, sharp, nourishing tidbits, first heard on the telephone, that were like a second meal in themselves and made Clara cluck with pleasure. "Darling," she said to Lester, through Florence's staccato beats. "Put on the Lehmann *Rosenkavalier*, yes? For me, please. For me and the boy." Babs and Stewie, sitting upright on their stiff rattan chairs, kept checking their watches. "Oh, God," Babs said, "just look at the sky. Hurry, before it goes." The mauve changed to pink as they all looked up, changed back to mauve, darkened in front of them. "Oh, God," Babs said. Alone in the corner of the sunroom, Adele sat reading a copy of yesterday's newspaper. She always sat alone in the Ottinger house. "Would it be so terrible if I went off?" Babs asked at last. Clara raised an eyebrow. "We just got up from the table," she said, needles clicking. The *Rosenkavalier* trio came on. "We both still have the Binswanger open house on the agenda," Stewie said. "Don't you worry about the Binswanger open house," Clara said without looking up from her crocheting. "It'll still be there an hour from now."

"That's a hundred-and-fifty points," Lester said sadly. Max shrugged magnanimously and looked pleased. Behind him, Manfred kept his silence. Max had already won a dollar and a quarter without his advice. He was doing all right. Lester was finished. A few minutes later, avoiding Clara's judgmental eye, Lester finally made a move to head upstairs for a nap, after paying up and quizzing Manfred about another aria. "Sing me the tune that goes with *Ah, fors è lui* . . ." He had trouble getting the words out, he was so sleepy. Manfred chirped a few notes, self-consciously, competing with *Rosenkavalier*, and reddened a little. "You know your stuff all right," Lester said. "Did you hear him, Clara? He really knows his stuff." Lester began to move off slowly. "You know how it is," he called to the family from the staircase. He waved to them all. "You'll forgive me this once."

Then Babs and Stewie left for the Binswanger's open house. "Take your own car," Clara said. "And send my love. Open house on Thanksgiving, of all days." Adele sighed with relief as her cousins moved off. "Well," Clara said. "Another year." "How many times do we say that every year?" Max asked. "Come over here," Clara said to Manfred. When he walked over to her, she leaned down from Olympus and kissed him on the forehead. "Surely the best Thanksgiving ever," she said with unexpected softness.

Manfred bowed his head respectfully. Clara's gesture had taken him off guard. As he stood there, he could hear the voice of Lotte Lehmann rise and fall back in the sunroom. She was singing in German. ". . . *hab mir's gelobt, Ihn lieb zu haben in der rechtigen Weis* . . ." What a strange language, he thought, how peculiar, how perverse, it was all backwards. Why hadn't he ever noticed that before? Everything German was peculiar and perverse. He had been staring at everything German all his life and had never recognized that. Outside now, it had suddenly become dark. Clara moved to kiss her brother, who waited for her dutifully. Manfred stepped aside to make room for them. His new cousin Babette had pinched his cheek in passing before she left the house and told him he was adorable. "Just adorable," she had said, whisking past him. It was her way, touch and run. She had other things on her mind. Still, the two words had flooded him with momentary happiness. His other cousin, Stewie, had shaken his hand, man to man. His palm was still warm where his pipe had been resting all day. Then he had scurried after Babs. He was crazy about his sister. He adored her. Manfred could tell. Manfred had an eye for those things. He always knew. Earlier, he had seen Rhea's left hand tremble uncontrollably while she helped to serve dinner; and it was clear that her husband, Shorty, his head bobbing slowly in front of his hump, just like a turtle's, was Uncle Lester's pal in drink. Manfred could tell; he knew. And there was more. There was always more. Florence Gordon was timid with her sister-in-law, Clara; it was like Manfred's shyness with Lester. More than timid, she fawned over her; she was fawning over Clara now, in the hallway, saying good-bye. It was the Ottinger money,

Manfred knew, it was always somebody's money. The Ottinger money turned Florence into a different woman.

"Watch out for the ice," Clara called, shivering on the front porch without a coat.

"Get inside, Clara," Max said, heading for the car. "You'll freeze to death."

"I'm all right. Now, remember to think of us as your second home, Manfred."

"Third," Adele said.

"For the last time, Adele, stand up straight," Florence hissed angrily, walking behind her daughter. "Look at your shoulders. We'll have to get a brace for you, is that what you want? Max, say something to her. Max?"

When they got home, Billy Brent was standing outside on Fairfax Road in the dark. Manfred could see him lounging casually against the Gordon sycamore, his hands in his pockets. His head almost reached the lowest branch. He wore his Indian sweater and a heavy muffler around his neck. He looked disapproving, as though somebody had just missed a pass. "I owe you a dime," he almost shouted at Adele, slipping something into her hand after Max and Florence had moved up the walk to the house. A flicker of concern passed across Adele's face. Manfred waited alongside her on the pavement. "It's not as though I'm going to China," she said to Billy. "I'll be here tomorrow." "I don't like to owe anybody money," he said in a voice that sounded false. "A-*dele*," Florence called from the porch, peering out at the street. "Manfred." Flat *A*. "They got you," Billy Brent said. "They got you both," and disappeared into the dark. Adele opened her hand. Manfred saw that she was holding a folded piece of paper. Even in the dark, Manfred could see that Adele's face was flushed. "Don't tell," she said, her voice suddenly full of energy. "And don't ask. It's nobody's business. He only pretends to owe me money. He gets this way. He gets excited, I guess. Billy's excitable." She slipped the paper into a pocket of her skirt and put a finger to her lips. She was beginning to look excited herself. Then she rushed upstairs to take a bath.

Another week passed, the strange aura became a little less strange. Manfred grew used to Florence's emotional spasms; they were like streaks of primitive crayon colors flying wildly across a blank page. She still fought Max whenever he brought up his ambitions for the business. Pay off the fixture loan, she said. Get out of debt. You'll mortgage the children's future, she added, putting her hand to her mouth melodramatically. But it seemed to Manfred that Florence meant something else, that she was really talking about herself, about her future, not the children's.

At the same time, Max became his friend. They began to play rummy together in the dining room after supper, keeping a record of their debts. Both liked to win, especially from the other. They had struck a deal. Manfred was now on the payroll at M. Gordon & Son at fifty cents an hour, exact working days to be decided. The money Max owed Manfred from rummy was added to Manfred's wages; it also worked in reverse; that way they pretty much stayed even.

In the afternoons, football went on. Manfred actually caught a pass and scored a goal. He was wearing his new saddle shoes at the time. Billy Brent smiled once and tapped him on the shoulder grudgingly. Joel Friedberg began to keep his elbows at his side, and Manfred moved into the backfield for good. In the bathroom late at night, while Adele made popping sounds with her bubble gum next door, he kept working at his new part, trying to make it perfect. As he worked, he thought about Adele. Sometimes she was friendly, sometimes she wasn't. Sometimes she sat in the back of the car with her leg touching Manfred's, sometimes she sat on the other side of the seat, far away, almost on the moon. At those times, it might have been that he didn't exist, that he was still somewhere in Europe, in Frankfurt-am-Main. He never knew what she was thinking, she always seemed to escape him. It made him wonder about her all the more. For Adele, words always came so easily, bullying chatter in the backfield, free-flowing advice at home, a kind of monologue without end; but she was careful about her secrets, that was where the line was drawn, it was something to respect. Manfred never intruded, never asked, although it was hard sometimes to restrain himself when they were washing the dishes together after supper. The Thanksgiving night episode with Billy Brent remained unexplained, and Manfred never caught her blowing kisses to the heavens again, or making bird sounds in the alleyway, either, even though he kept daily watch from his bedroom. Adele took the garbage out, lifted the lid to the can, dropped the bag, and turned quickly to run back to the house. That was the way it went every night when it came time for her chores.

Chapter Six

He started school a week later, accompanied by Florence, who had put her hair up in curlers the night before for the occasion, and Adele, who sat calmly picking lint off Manfred's sweater all through breakfast, saying hardly a word, just pursing her lips a little to help her concentrate. Ambling along a couple of yards behind them, up Oakfield Avenue, were Billy Brent, trying to stay awake, Jojo McAllister, Laura Piscitelli, and silent Kitty Dean, who lived down the street next door to Billy. Florence walked with quick, hungry steps, greedily chewing up space on the sidewalk. She never ambled, especially when she was on her way to deal with Fate. At her side marched her children, striding along with her. Ahead lay The Future. How she dealt with Fate would decide The Future. Nervous energy crackled around her like shafts of lightning. By the time they left the house, Florence had cooked some doughy oatmeal, washed the bathtub, uncurled her hair, and rinsed out two pairs of stockings. "You almost ready?" she kept shouting at Manfred from the kitchen, while Max, stuffing down the oatmeal under thick layers of butter, tried to stay out of the way. "Can't be late today," was all he had to say, in a patient voice. "For God's sake, keep your hands off him," Florence yelled at her daughter, who was blowing a faint wisp of lint off Manfred's right shoulder. "Just don't forget to smile," Adele was counseling him. "Everybody loves a nice smile." Meanwhile, Manfred's breakfast had already begun to rise in his throat; he swallowed hard and thought of the blue vial upstairs; then he thought again; it would not do to pass out in a delirium on the steps of Forest Park High on his very first day there.

Every now and then on the walk, Jojo McAllister's whinny rose in the air behind them, a terrible sound. Laura Piscitelli shrieked something incomprehensible. They were both laughing at Billy Brent, who had said something under his breath, meant for them alone. It was not normal laughter.

There was something agitated running through it, something wild and out of control. It was like screaming. At the sound, Adele grinned and glanced across at Manfred. For the moment, flashing her smile, she looked pretty wild herself. They must be talking dirty again, he thought. It was the only thing that made them sound like that. They were always talking dirty, all except Adele, even in the middle of touch football, even while Billy Brent's violet eyes were trying to hold Adele's in a single-minded, proprietary bond. Maybe they were looking at one of those little beat-up books, another choice volume from Billy's unusual collection, on which Adele always, or almost always, turned her back, without seeming prissy or righteous (Manfred admired that). When Billy finally let Manfred have a look during touch time-out, just once, with Laura Piscitelli's special permission, Manfred had discovered his favorite Gumps with their pants pulled down to their ankles or their skirts hiked up to their chins. A glimpse was enough to make him think he was going blind. Bim Gump was doing terrible things to his sister-in-law, Min, Min was doing terrible things in return, while Chester Gump, her son, watched them both.

"Quit dawdling now," Florence ordered when they reached Liberty Heights Avenue. "It's five of nine."

"Golly-gee, kids, it's five of nine," Laura Piscitelli squealed in an affected voice that exactly caught Florence's intonation.

Florence ignored her—actually, they rather liked each other—and doggedly headed for school. Then Manfred, glancing over his shoulder, saw Laura cross her eyes and pretend to walk bowlegged. Jojo whinnied again. Heh-heh, he brayed, slapping Laura on the back. Head down, Billy Brent paid no attention to them. He was stumbling along, leafing through a textbook, his rosebud mouth tense with concentration. Little Kitty Dean, alongside them but apart, walked a straight line on the curb, as she always did. It was like being trailed by the enemy, Manfred thought, up front. It was like being totally exposed. You had to keep an eye out for trouble when you turned your back on Billy and Laura and Jojo McAllister. They had nerve. It was one of the first things Manfred had noticed on Fairfax Road. Nothing seemed to scare these kids, even the shy ones like Kitty Dean. Once they made up their minds, once they were in motion, you could almost see the earth slide away from under their feet, everything went so fast.

Mr. Lund, the principal, allowed Florence only five minutes. "Thank you, thank you," he kept saying, as he maneuvered her towards the door of his office.

"Isn't there anything else you want to know?" she asked in a disappointed voice. Her hand went to her hair, patting the brittle set as though to console it for having been put to so much trouble for so little. "I have duplicates of the boy's papers and everything."

"I'm sure Miss Kleeman has all the papers she needs," Mr. Lund said, nodding affirmatively at his secretary.

"Yes, sir," Miss Kleeman said, tucking Manfred's skimpy file under her arm.

Florence was beginning to lose her expectant look. Fate was slipping away.

"There's nothing to worry about," Mr. Lund said. His pants were baggy, his hair grey. The walls in his office were also grey. A grey portrait of George Washington hung on the wall. Beneath it, Manfred sat on a black chair, biting his nails. "We have a standard procedure here for such matters," Mr. Lund went on. "We are not without experience, of course. The focus, as it should be, is on the pupil and solely on the pupil." He gazed at Florence meaningfully.

"I just thought," Florence said, "given the special circumstances . . ."

"All circumstances are special," Mr. Lund came back. "All factors are unique."

What could she say?

"The whole point of the next couple of weeks," the principal continued, "is to offer the young man a chance to become acquainted with us and for us to return the favor. The future will take care of itself."

Manfred began to make a clicking noise with his teeth.

"Well, at least let me leave him some lunch money. I thought we'd be back home in time for lunch. I didn't think he'd jump in so fast. Are you sure . . ."

"Leave it to us," Mr. Lund said. "We know our business. We are professionals. There is no need for concern, I assure you." He turned to Miss Kleeman. "Would you ask Miss Ficker to please come in? Now," he added, "thank you very much, Mrs. Gordon. I hope we'll see you at parent-teachers. That would be very helpful for all of us. For the rest, there's no need to disturb yourself any further. Manfred is now ours. In this building, he belongs to us." Mr. Lund smiled to take the edge off his remarks.

There was a grey blotter on Mr. Lund's desk. His tie was also grey. He continued to smile absently at Florence, then at Manfred. Manfred's new coat lay in his lap, he was playing with the collar. Florence shrugged. Face to face with bureaucratic destiny, she sometimes became intimidated. "I guess it's out of my hands," she said, giving Manfred a bemused look. "Don't forget to tell them how many languages you know. Don't be shy. Speak up. He does know a lot of languages. I think he's a year, maybe two years, ahead of himself. That's what they say. I have other papers from Frankfurt . . ."

"Rest assured," Mr. Lund said vaguely.

A tall, slant-jawed woman entered the office. "You called for me?" she asked.

"Ah, good," Mr. Lund said. "This is Manfred Vogel. And Mrs. Gordon. You remember our discussion about Manfred Vogel, I'm sure." He held Florence's coat for her, moving her along. Manfred got to his feet. The woman looked at him. "I certainly do remember," she said. "We need to have a chat, you and I. We need a chance to get to know each other."

"Take another fifty cents," Florence said, holding out the change to Manfred.

"I have enough," he said.

"Take it," she ordered.

"I'll see you later," Manfred said.

Florence was already on her way out of the building when Manfred found himself seated in another office, facing the tall woman across the desk.

"Well . . ." she began, folding her hands in front of her, as though she was the best behaved girl in the class.

Her name was Miss Ficker, she told him; then, after letting it sink in, she spelled it out in a slow, patient voice. Miss Ficker lisped. She also wore thick glasses that almost hid her eyes. But so did Herr Levi in Frankfurt. It seemed to go with teaching, with having a certain kind of brain. Manfred gazed at Miss Ficker comfortably. He was looking for signs of susceptibility. It was always easier for him with women than men teachers. Women teachers took to him, had more patience, sometimes they even liked to touch him. Had he met with Mr. Burns? Miss Ficker asked. Unusual. He really should have met with Mr. Burns first. Miss Ficker was assistant principal for girls, Mr. Burns was assistant principal for boys. Perhaps Mr. Burns was in study hall. There always seemed to be a problem in study hall. "Now," Miss Ficker went on, opening Manfred's folder. Then she checked his records approvingly—geometry, ancient and European history, French, English, Hebrew—described his new school in elaborate detail, and set forth his new courses. He would take German, Miss Ficker said, to help perk up the class. "Any questions?" she asked. There were no questions.

"Perhaps you'd like to try Miss Klinesmith. Miss Klinesmith is Alsatian. She teaches French. I will put you in the hands of a Forester today so you won't get lost making your way around. Foresters are our student monitors. A high honor, indeed. Come now, don't poke, you've dropped your coat again . . . Now, as to your Forester. Miss Kleeman," Miss Ficker called, "would you get me William Brent, please?"

On the way to Miss Klinesmith's classroom, on the third floor, Billy looked riled. "I didn't ask for you," Manfred said bravely. "It was her idea."

"I'm supposed to be kicking a soccer ball around this period," Billy said. He yawned ostentatiously behind his hand.

"You'll hate me," Manfred said.

Billy laughed sardonically. "I don't like to cut soccer."

"Don't hate me."

"I only get to play soccer twice a week."

Manfred stopped on the staircase. "Go play soccer," he said.

"Come on," Billy answered, waiting on the step above Manfred.

"You go play soccer. I'll find my own way."

"I don't take orders from you. You take orders from me." Billy grabbed Manfred's arm.

"Don't touch me."

"Holy Christ, you're stubborn. You *never* listen."

"Apologize."

"Get a move on. I don't apologize to you. And who the hell are you to order me around? What a pain in the ass you are. From the beginning."

"I scored twice the other day."

"With my passes."

"So?"

"Anybody could've caught them. Anybody did. Are you going to move or aren't you?" Billy was now beginning to look really threatening. His mouth was clamped, his face pink. Manfred decided to move. "I'll leave you in Klinesmith's class," Billy said through his teeth, "and pick you up at the end of the period. I've got you on my hands all day. The *whole* friggin' day."

After this exchange, they were two minutes late to Miss Klinesmith's class. Another minute was wasted as Forester William Brent explained his mission and turned over a note from Miss Ficker. Miss Klinesmith read the note, passed it back to Billy, and looked sharply at Manfred. "Vogel?" she asked. "Yes." "German for bird." "Yes." *"Alors,"* she went on, *"asseyez-vous. Vous comprenez?"* She pointed to an empty desk. *"Oui, merci,"* Manfred said, feeling the class stir behind him. Miss Klinesmith came up to Manfred's shoulders. She had a tiny head, cropped white hair, and a little beak of a nose. *"Eh, alors,"* she said, turning to Billy Brent. "You'll come for Monsieur Vogel at the end of the period, not a minute before, not a minute after. On your way now. Now, *mes élèves*," she began, turning to the class as Billy whisked out the door. "Will all the absentees please raise their hands?" Miss Klinesmith was famous for such remarks.

The class proceeded to its work, birdlike. They pecked at the seeds of elementary French together, conjugated a few essential verbs on the blackboard, stealing from each other like magpies, translated a couple of classic French fables in singsong voices. *"Très bien,"* Miss Klinesmith said every time Manfred spoke up. He soon discovered that he couldn't speak without the color rising to his face. There were twenty-one other pupils in the class, fourteen girls and seven boys, all sitting behind him. If he could hear his glottal Hessian accent ricocheting off the wall of the classroom, what did it sound like to them? He didn't dare turn around; they must loathe him. Stranger, latecomer, parasite, goody-two-shoes, and too smart, in the bargain. "Well, Monsieur Vogel," Miss Klinesmith said, making matters worse, "it seems that you are too advanced for my poor beginners here. Yes, ladies and gentlemen?" *"Oui,"* a few girls called out listlessly. *"Eh, bien,"* she said, perching herself on her desk, right above Manfred. "I would like to say a few words now about the past tense. We will start with the verb *être*. Everything starts with the verb *être*. Now . . ."

"Isn't that so, Monsieur Vogel?" Miss Klinesmith said from above a few minutes later, showering Manfred with spit. He looked up at her with sleepy eyes. The room was very warm. *"Oui ou non?" "Oui,"* he guessed. *"Très bien,"* she said happily, clapping her hands.

The four-minute bell rang. The class began to stir again. A few girls began to put on makeup. The seven boys, all of whom sat in the same row, began to talk among themselves. *"Eh, enfin,"* Miss Klinesmith said, summing up just as the regular bell sounded. Within a few seconds, Billy Brent was at the door. The class shuffled out, eyeing Manfred solemnly. Billy waited outside. "You know," Miss Klinesmith said, "I am from Guebwiller. You know Guebwiller, near Sassenheim? Alsace," she went on in a rapturous voice, "the most beautiful province in all France! And the most beautiful province in Germany, too! I assume you understand what I mean."

"It's getting late," Billy Brent droned from the corridor, pointing to his watch. "He has another class. So do I."

"Allons," Miss Klinesmith said. *"Jusqu'à demain."*

"How was soccer?" Manfred asked, as they walked down the stairs.

"I sat in on music instead."

"You sat in on music?"

"Quit repeating everything I say."

"You said you wanted to play soccer."

"Well, it got too late," Billy said in an accusing voice. "It was too late to change my clothes. And I like music," he added belligerently. "I like to sing. I sing tenor in the Boys' Chorus."

"What is Boys' Chorus?"

"This way, sonny boy," Billy said, turning a corner. "Now, listen." He then began to sing in a high, light voice. " '*Give* me some *men*, who are *stouthearted men*, who will *fight* for the *rights* they *adore* . . .' I can get you in if you want. Not everybody can get into Chorus."

"I don't mind."

"Say yes or no."

"Yes."

"You're in. You carry a tune?"

"I can sing opera. I mean I know some of the melodies."

"We'll see about that," Billy said, looking dubious. "Miss Tubman knows the real thing when she hears it . . . 'and we *stride* as we *go* off to *war*,' " Billy Brent sang out. "Shit, we're here," he said, finally sounding cheerful.

"Where?" Manfred asked.

"American History, Miss Feinstein. A real sob sister. Now listen. Adele has study hall next door this period. You give her this note at the end of the period." He pushed a folded piece of paper at Manfred. "Just wait for her right over there. Then give her that."

"Why can't you give your own notes to Adele?"

"Do me a simple favor."

"It's stupid."

"Now listen, Fritzie."

"You give your own notes to Adele. You write them all the time."

Billy folded his arms. "I just want to make sure she gets that note, in case I miss her. Got it? Jesus, what a pain."

Miss Feinstein opened the door to the classroom. "Boys," she said sadly, dabbing at her eyes with a moist handkerchief.

"Manfred Vogel, ma'am," Billy said. Then he passed along Miss Ficker's message. "I'll meet you here after class," Billy said to Manfred, giving him a good hard look. *"Thanks-for-the-favor."*

Manfred stuffed Billy's note into his pocket and followed Miss Feinstein into the classroom. Thirty-four students sat inside. Half of them were slumped over their desks, bent double, the other half were doodling in their notebooks. Manfred sat there for forty-five minutes, staring out the window, hardly understanding a word that was said. Miss Feinstein spoke in a monotone; she feared a student uprising so she never challenged. She was talking to the class about the Spanish-American war and had trouble filling out the period. There was not enough drama once the Maine was sunk. Teddy Roosevelt, she said, sighing. Roughriders. Manifest destiny. And so on, including the charge up San Juan Hill, twice. There was a snore from the last row. Two girls played tic-tac-toe. The doodling continued; sultry daydreaming. Under the giant windows, the radiators knocked and rattled. Miss Feinstein's eyes ran; she blew her nose; her sweet, round, passive face filled with pain. The four-minute bell rang, electrifying them all.

Outside Miss Feinstein's classroom, Manfred swept the corridor with a long, careful look, trying to find Adele, hoping to miss her. He wanted to throw Billy's note into a wastebasket; let him run his own errands. Girls were pouring out of the study hall next door. Beyond them, teachers monitored the intersections. Foresters wandered along, wearing armbands, looking for student criminals. Adele came walking at him, smiling her rich plump smile. She was wearing her new loafers with the pennies in them and her purple sweater. "There you are," she said brightly, as though she were expecting him. Adele carried her schoolbooks clasped to her chest, embracing them with both arms like a lover. "I've been thinking about you all morning. I kept looking for you when classes changed, but I kept missing you. Who do you have for French? Did you listen to me about smiling? Why aren't you smiling now?"

"I don't like this," he said, reaching into his pocket.

"Give it a chance," she answered. "It's only the first day. It gets worse."

"Not school," he said. "That's not what I mean." Their eyes locked. Adele waited. "Why should I have to deliver Billy Brent's messages?" he asked.

"What messages?"

Manfred held up the folded piece of paper.

"What-are-you-talking-about-explain-to-me-please," she said, in a flat voice.

"This," he said, waving the paper.

"That's from Billy?"

"Yes."

"I should have guessed. Well, don't ask *me*. Ask *him*. I don't know anything about it."

"I won't do it again. Just because he asks."

"I told you, I don't know anything about it. It's between you and Billy."

"Well, I won't do it. Just remember that."

"Oh, don't get so frazzled," Adele said, sounding exasperated. "You have a tendency sometimes."

"Speak English."

"Look who's talking. Get off your high horse, Manfred. It's only a piece of paper."

"He must be getting excited again," Manfred said, primly. "I hear that he's very excitable."

"Why, you little . . ."

"Here," he said, thrusting Billy's note into her hand; and he was off.

"I won't forget that," Billy said, sidling up to him out of nowhere. "I don't forget pals." "Pals," Manfred said, looking at the floor. Billy reached over and patted him on the buttocks. "Don't try to act so tough," he said. "Save it. We don't want to spoil Boys' Chorus, do we? I mean you want in, don't you?"

Manfred finished off the morning with Bigprick Lambrino. Bigprick Lambrino, who was only twenty-six, slunk around the classroom, pelvis out, shoulders rounded, drilling his troops through German exercises, as though they were out on battle maneuvers. His hair was slicked down front to back. *Der*, he shot out, repeating himself like a machine gun. *Das. Die.* Anybody who confused those three little syllables got the first nail in his coffin. Manfred's eyes popped. Now, Bigprick went on, slithering around the room like a garden snake, repeat after me, all of you, John McAllister included, John McAllister especially . . . John McAllister sat next to Manfred. He was hopeless. He couldn't learn *der, das, die*. He couldn't pronounce umlauts. Already, he had two nails in his coffin. Two more nails and he would be six feet under. John McAllister looked down at his desk top. His pug nose twitched. The elastic on his corduroy knickers was torn, one sock was halfway down the calf of his left leg. Manfred whispered an answer to him. "That will be quite enough, Herr Vogel. Cheating is a serious matter around these parts. I don't know about your parts. You don't believe me? Wipe that grin off your face. Get rid of that awful smile. What *are* you grinning at?"

Finally he made it to lunch. The rest was easy, more or less. Billy kept him in tow, sat Manfred and Jojo McAllister at his own table in the cafeteria with his own classmates, senior giants. There were four of them, and they were talking dirty again. Somebody passed around a pack of condoms for a quick inspection, and Jojo McAllister, famished from his ordeals in German class, ate three chocolate bars. The stink of hard-boiled eggs was everywhere, cheap mayonnaise, rotting apples. Everyone was shouting. Dishes banged, feet stamped, a whistle blew. Through the noise, Manfred said nothing. He sat quietly, barely moving. He did not want to appear to cling. He was terrified of appearing to cling. He would stay apart from the four senior giants and especially from Billy Brent. He hated people who stuck to you like burrs when they weren't wanted.

A little later, during the break that followed lunch, they all headed outside for some fresh air and exercise. Jojo stayed at Manfred's side, complaining about Bigprick, while Billy Brent mumbled an excuse and headed towards the cinder running-track, where Manfred could see Adele standing alone.

Billy and Adele began to walk around the track together, deep in conversation. Behind Manfred, the school stood in a lush pile of graceful pink brick, laced with Tudor flourishes. There was an auditorium off to one side, a separate shop building, two gyms. The campus seemed huge, it spread in every direction. By now, Adele and Billy were heading towards the far curve of the track, a block away. Somebody raced up to Manfred and flipped his tie out of his sweater. "Cut it out, Goddammit," Jojo yelled. "He's a new boy." It was cold standing there, snow threatened again.

From where they stood, Manfred could see Adele's coppery hair shining in the distant grey light. Her bright yellow scarf, the color of a buttercup, flashed beneath her smile as she and Billy started on their way back. Walking along, kicking cinders in front of her, Adele kept pulling away from Billy's touch. "Hi, sweetie," a familiar voice called out, sounding affected. It was Laura Piscitelli, off to one side, beating herself to keep warm. Three boys surrounded her. "I'm talking to *you*," she said, squinting at Manfred threateningly. Manfred looked away and pretended not to hear. "I mean Lambrino's out to bury me," Jojo was saying, single-mindedly. "Why me? What did I do?" Out on the track, Billy was waving his arms, trying to make a point.

Adele's scarf flashed again. Billy Brent was coughing and gesticulating at the same time. Laura Piscitelli, hands cupped around her mouth, yelled something at Manfred and Jojo. "Ignore her," Jojo said. "She never knows when to stop. Don't pay any attention." He glanced at Manfred, then made a rude gesture with the forefinger of one hand and the closed palm of the other. "Get it?" Jojo asked, nodding in Laura's direction. "Ah, so," Manfred said stiffly, in his German way. Laura was squealing something incomprehensible now and laughing. "Oh, shut up," Jojo muttered. Up ahead, Adele and Billy were still making their way around the track, arguing.

Chapter Seven

Whenever M. Gordon & Son happened to be working together behind the counter, it was clearly impossible for them to move at the same time. As Adele had said: two by four and, she might have added, barely room to breathe. The store was too small. They jostled each other, bumped hips, ricocheted off the sharp countertop to avoid colliding. "Pardon," Manfred said, pulling back against the wall. "Sorry," Max responded. "I'll try to stay out of your way," Manfred said. "It would help," Max answered. A skinny display-case ran the length of the store, containing watches, bracelets, earrings, pins, thin gold necklaces, school rings. Max was hoping to corner the market in Baltimore for school rings. That was one of his "plans." If he could own every senior class in town, he told them, he would be able to sell about three thousand rings every spring in time for graduation, and it would be clear profit, no inventory. Naturally, there were problems. A jeweler over near Patterson Park, a dumb Polack, had the high schools in south and east Baltimore sewn up. Max thought it was a fix, everybody said so, but fix or not, that still left City, Poly, Western, and Forest Park, where there were plenty of rich kids for the pickings. Not a bad haul, if he could pull it off. Did it need a fix? The question bothered Max. But he kept putting off the big push. He was still planning it. He liked to plan. It nourished his optimism. Soon, he thought. He hated the idea of a fix.

Manfred ran errands for the business downtown: deposits to the bank on Redwood Street, for which he had to show a special identification card, a delivery now and then to fancy offices off Charles Street, pickups from the wholesalers near the docks, where the prows of foreign ships loomed over Pratt Street like soaring steel birds, sometimes an excursion to buy something for Florence at one of the cut-rate department stores nearby. There were stockings on sale at Gutman's. She wanted a silk slip at the May Company that was half-price, an apron, dish towels. Once she said she needed a brassiere, 32-A, she said, flapping a sheet in the air as she made the bed in the passion den. As she spoke, Manfred turned his face to the wall, staring at the pink nymphet over the bed and humming *Rose Marie* under his breath, trying to blot out her words. Brassieres! On Pratt Street, the ships rode high in the water, their prows thrust at the city like daggers. They all had strange names spelled out in clear letters up near their top decks. *Merseyside* from Liverpool, *L'Espoir* from Cherbourg, and, not so strange, the *Adalbert* from Lubeck. They came up the Chesapeake side by side, sailing north along the great roadway, funnels smoking, then waited their turn down near Annapolis to enter the port. Once in place at the docks, they sat complacently alongside each other, as if they owned the town. Pratt Street was one of the places Manfred liked to be.

Running around the streets on his own, he began to learn a little about the city. North Avenue divided it in half, uptown from down. Charles Street did the same, east from west. It was better to live north than south, west than east. South meant negroes, barricaded in a tight little city of their own, east meant tough Catholics from Poland who worked in the steel mills. Both were dangerous. North was where the money went, west was full of aspirations; Fairfax Road ran exactly between the two. As for M. Gordon & Son, it lay downtown, just west of Charles Street, on Lexington, near the Loew's Century and its upstairs neighbor, the Valencia, where clouds floated across the ceiling during the movie. Across the street was O'Neill's Department Store, in which Jews, Max warned his colleague, neither worked nor shopped, and just a few feet away was Huyler's, home of chocolate sundaes and overpriced ice-cream sodas. There were a lot of movies around, a vaudeville house called the Hippodrome, a cluster of other department stores, and a lot of cheap specialty shops. It was not like Frankfurt, where the whole city was downtown. It was not like anything in Germany. Here in Baltimore a kind of slow commercial fever burned daily within a tiny, compressed area known as "in-town," and the rest of the city, lazily stretching to the north and other parts, like an incurable rash, went its isolated, patchy way.

Max had a membership in Hagenzack's Health Club, which occupied the whole tenth floor in a building on the corner of Lexington and Charles. "My only extravagance," he always explained, before adding, "I'm sorry to say." Like everyone else, Max decided to join Hagenzack's out of pure vanity. There was a growing pot below his navel, as he demonstrated to Manfred one afternoon while touching his toes ten times. "Watch gravity do its horrible work," he breathed, as his stomach fell into his pelvis. Bending over from the waist like that was certainly no help. Manfred could see that by the time he reached the count of five, aiming for Hagenzack's floor, he was also purple in the face. Max's vanity: he gave his beautiful white hair a hundred brushstrokes every night, then enclosed it in a net when he went to sleep, for safekeeping. Nothing marred it then. Nothing disturbed it. In the morning, at breakfast, it glistened in impeccable waves. Max knew how handsome the contrast was between his honey skin and his white hair. Enough people had commented on it over the years, and all he had to do in case he forgot was look in the mirror, which he did, perhaps, a little too often.

Max also liked Hagenzack's for the company it provided; so did Manfred. Uncle Lester Ottinger was a member. A half-dozen city judges who lived out in Forest Park or around Lake Drive worked out regularly on the aging equipment. There were old cronies from Max's childhood, a few strange faces here and there that might turn out to be useful business contacts, and even a rabbi from one of the reformed temples, whose massive, hairy body always shocked Manfred when he saw it heaving away at a barbell. The place was convivial. Everyone looked the same, more or less, without their clothes on, everyone talked to each other whether they were introduced or not. Max would chat in his jockstrap with a total stranger and discover later, from Lester usually, that he had been in conversation with a member of the Board of Education. How easy it appeared, how natural. One time it was a friendly test pilot from Glenn L. Martin, trying to stay in shape, another a Baltimore city councilman. It seemed important to be part of it. It created opportunities, Lester said, it broadened the horizon. Meanwhile, Manfred wandered alone from one corner of the club to the other, waiting for Max to finish up. He wasn't allowed to participate because he was underage. He could only observe. Old Man Hagenzack condescended to slip him a word or two in elementary German, in passing, and that was that. You had to be twenty-one to be a member of Hagenzack's. It was the house rule. Manfred had discovered that twenty-one was the age you waited for in America. Before that, you hardly existed outside yourself.

Lester always showed up at Hagenzack's on Wednesday afternoons, one of Manfred's after-school workdays. He dropped Clara off for some shopping and a matinee at Ford's Theater, had lunch with his lawyer or his broker, then joined Max and Manfred up on the tenth floor, after an hour or two spent at the main library on Cathedral Street, reading the latest periodicals from abroad, French, German, and English. Lester worked out barechested. He had hair on his shoulders and upper back, ugly wiry stuff, his arms were as stringy as Max's, and sometimes, as he strained at a weight or jumped rope, his errant eye would suddenly start on one of its independent journeys. He groaned as he worked, complaining to Max alongside him, complaining to Manfred. His eye drifted slowly to the right, then drifted slowly back. Tiny veins on the side of his nose shone pink. Soon, he was panting. "You have to be crazy," he cried out to Manfred in a wild voice, the tendons stiffening in his neck as he struggled unsuccessfully to lift the weight. "Why am I killing myself like this?" It was well-known in the family that Lester rarely shouted; neither did Max.

The place smelled of dry sweat and bourbon; the voices were thick and unguarded, the vocabulary pungent, even when the rabbi was around. Old Man Hagenzack, who was seventy-two (some said seventy-eight) and still a rooster, was always on call, ready to offer professional encouragement to his clients, who needed it on a regular basis. He strode around the club in his ankle-high sneakers, with a whistle hanging from his neck. Deftly, he jumped a mat, straddled a body, tested equipment. When he swung on the parallel bars, his aging dewlaps shook. As he demonstrated push-ups, his bony head glistened under the bright lights. He wore white ducks and a sweatshirt. He was the master, they were the slaves. "You're really beginning to look like something, Lou," he said, examining a huge pile flattened on the floor. His voice was thick with Central European echoes; Old Man Hagenzack had been born in Hamburg. "Chust give it one more try. Come on, it won't bite. Not like that. Chin only. Look at me. Stiff, hard." Then he walked over to Max. "You're not looking so bad, either," he said, sounding a little skeptical. "I call it progress. Yes, it is better. Now, let me have a look at the waist. Pull up the shirt a minute. Hard. Firm. That's the ticket." And he was off, to check the towels. He fondled everybody as he flattered them, Max, Lester, the whole gang, stroking their upper backs, poking their stomachs to make a point, holding their hands in a grip that could hurt, that actually made them wince. Hard, firm, stiff; Old Man Hagenzack had muscles and he liked to show them off. While he mixed in democratically—at Hagenzack's all slaves were equal—doing a little worthy public relations where it counted, pressing sweaty flesh and leaving his fingerprints on everybody, the members shuffled from one activity to another, taking their own good time about it. They set the pace, and they would not be hurried. Slaves or not, they paid the bills. The members wore straw sandals on their feet, provided by the management. Some of them wrapped themselves in damp sheets, to keep cool. A few secure souls, pretending to be indifferent to everyone else, strolled around naked, chests out, although Old Man Hagenzack disapproved of such exhibitionists; he liked jockstraps on his clientele, to prevent hernia. Mounds of identical freed genitals swung in the air in the shower-room, a couple of feet below Manfred's line of vision, black forests of steaming pubic hair, additional thickets tufting the lower back, matting the buttocks, or running in a light, thin line up the cleft between the cheeks. Like muscle, hair was one of man's flourishing provinces, sometimes in unlikely places; that much was clear at Hagenzack's Health Club.

No doubt, it was serious business. Without question, it was hard work. It was the hardest work any of them did, building bricks out of nothing. It wore them out, occasionally made them feel faint from the strain. It was also hopeless, but no one talked about that. It was the effort that counted, the impulse, the continuing display of manly will. That added to the communal feeling. Together they shared an ideal of perfection. Old Man Hagenzack merely exploited it. Cocks up, gentlemen! the friendly test pilot said one day, as they all waited their turn at the chest-expander. Dicks high! That was the real point, after all, potency, potency forever, you didn't have to be brilliant to understand that. The language might be a little loose for Max and Lester, a little racy, perhaps, but they all laughed together, self-consciously, Max maybe loudest of all.

Most Wednesdays, after Hagenzack's, Lester came back to the store with them. While they were gone, Max simply put up an old sign on the front door that read "Back in an Hour." That took nerve, Manfred thought, to close down the store for an hour in the middle of the week. He didn't know anybody else who did that. It was Max's convenient theory that people who were looking for luxury merchandise would always come back if they couldn't find it the first time around. To help them, he stayed open an extra hour on Thursday night.

With the three of them standing side by side in the narrow shop, late in the afternoon, much quiet courtesy was required. It was just too crowded.

When a customer showed up, Lester disappeared into the back room, and Manfred found another errand to run. Then they would get back together ten minutes later, as soon as the coast was clear, and continue to make small talk. (They were an odd trio, but they liked each other; and who else did they have?) Once a week, for three weeks, it was Manfred's job to check the results of the Schick ad, which ran on consecutive Sundays, and fill all the orders. To date, there had been forty-seven returns. Not bad, Max said, flipping through the coupons; not wonderful, either; at least something, including promotion for the store. Make a roster of the names, he told Manfred, so we can start a mailing list. Then he held up a specially printed sign, the letters running across it in a large, flowing script. "Marriages are made in heaven," it read, "but engagements are made at M. Gordon & Son. Time payments acceptable."

"What do you think?" he asked Manfred and Lester.

"It's catchy," Lester said, his head tilted critically to one side. "But it takes some time to figure out. I mean it's not instantaneous like 'good to the last drop.'"

"What does it mean?" Manfred asked.

"It's an old saying we have in America," Max said, "stolen, adopted, and slightly transformed for my purposes. Maybe repetition will do the trick." He stuck the sign in the window, arranged some fake engagement rings around it and a few wedding bands.

That Wednesday night, by prearrangement, Lester was planning to take Manfred and Max out for an evening on the town. He had been planning it for a week. Clara would have to find her own way home, get a lift out to Mt. Washington with one of the old acquaintances she'd pick up at Ford's after the matinee performance of *The Constant Wife*. The men would take in an early movie, then drive uptown in Lester's smart, new-model convertible, grab a bite at Nate and Leon's on North Avenue, and head for home. "You're my guests, of course," Lester said, half expecting polite objections from Max, and finding none. "Well,"—sighing—"what do you guys want to see?"

Shirley Temple was at the New. *A Tale of Two Cities* was up the street at the Century. Lily Pons was at the Hippodrome, in her first picture. "Would you like to see Lily Pons?" Lester asked Manfred. Manfred wanted to see *Captain Blood*, which was playing at the Stanley. "Ah, *Captain Blood*," Lester said, looking serious for a moment. "But before you make up your mind," he added, "you should know that the Disappearing Water Ballet is also at the Hipp. It's the main attraction in the stage show." "The disappearing what?" Max asked. "Water Ballet. Sixteen broads," Lester said out of the side of his mouth, "who dive into a portable swimming pool and disappear in front of your eyes, don't ask me how or why. I only hope they can swim." "I'd like to see *Captain Blood*," Manfred said, putting merchandise away. Max, on the other hand, leaned to *A Tale of Two Cities*. He liked Dickens. It was convenient, too, he reminded them, just a couple of yards up the street. "It got the Blue Ribbon with Palms in the *Evening Sun*," he said. "So did *Captain Blood*," Manfred answered, slamming a drawer shut. "And if worse comes to worse," Max added slyly, "we could always watch Marvin Manning's organ rise." He was referring to the powdered instrumentalist who surfaced in the orchestra pit between shows at the Century and played stirring music on his mighty Wurlitzer. "You sure now you don't want to see Shirley Temple?" Lester asked. There was no answer. "Well,"—rubbing his ear thoughtfully—"between you and me, I promised Clara I would take the boy to see Lily Pons. She insisted." "I hate coloraturas," Manfred said. He also hated to be called "the boy." "Well, what about the water ballet?" Lester asked. "I'd rather see *Captain Blood*."

Lester threw up his hands. The negotiation was over. Clara had lost. But he had fought the good fight and played by the rules; and he wanted to see *Captain Blood*, too. He began to give Manfred and Max a hand with the merchandise. Tomorrow was the bookkeeper's day, everything had to be in order. They passed boxes between them, hid them away, made everything neat. Max tested the nightlight, switched on the burglar alarm, turned the locks on the door, admired the new sign in the window. "Marriages are made in heaven . . . " Then off they went. Driving the half-mile or so uptown to the Stanley, they stuffed Manfred crosswise into the tiny back space in Lester's ace, new-model convertible, where he sat with his knees scraping his chin and his arms hugging his knees, suffering leg cramps.

A few hours later, over the brisket, Lester said, "Just tell Florence we saw Lily Pons." He had dropped them off at Nates after the movie, parked the car, and joined them in a back booth ten minutes later, looking flushed after a quick shot in a bar across the street.

"Why should I tell her anything?" Max asked.

"You'll have to say something. She'll want to know where we were tonight. That goes for you, too, Manfred."

"I know how to lie."

"There are lies and there are lies. Lying's nothing to be proud of, speaking in general. Just say Lily Pons. And water ballet. Describe the water ballet. They dive in and disappear. It's a trick with mirrors. It has to be. And don't eat so fast. You're gulping your food down."

"I could tell those were false ships in the movie," Manfred said, after a moment. "They looked like stick toys to me."

"It's what they call special effects," Lester explained. "They have these tanks in the studio and they fill them with little ship models. They're perfectly scaled. I saw pictures in the *Sun* once. Then they turn on the fans. You can make a typhoon like that. They can do anything out there."

"The whole thing cost a million dollars," Manfred said. "I read it somewhere."

"Publicity," Max said, chewing carefully on the brisket. He had passed all his potatoes along to Manfred. "And while we're on the subject," he went on, "I'm thinking of doing a little publicity of my own, advertising on the radio."

"That should cost a pretty penny," Lester said. "How much?"

"Enough."

"What's enough?"

"I like that line I dreamed up," Max said, avoiding Lester's question. "I want to use it."

Lester looked doubtful.

"Why not?" Max asked. "All it takes is repetition. That's the secret."

"Marriages are made in heaven," Lester intoned. "I still think it's hard to remember."

"But engagements are made at M. Gordon & Son. That's the important part."

Lester still looked doubtful.

"I deal in beautiful things," Max said. "People know me for that. Why shouldn't I cash in?"

"You certainly don't need my permission to advertise on the radio," Lester said. "Who am I to complain? Hard to figure money these days, anyway. I just hope radio isn't too expensive. Never overextend. Isn't that the rule?"

"When people think jewelry in Baltimore, Maryland, I want them to think Gordon. Like when they think drugstore, they think Reads."

"Run right to Reads," Manfred said.

"As the saying goes. And they do, by the thousands."

"Well, good luck," his brother-in-law said.

"I figure I have as much luck coming as anybody else. The rest is smarts. And a belief in the future." He turned to Manfred as he said this. He was smiling. He had been smiling all through the exchange.

"And what do you hear from Deutschland, Manfred?" Lester asked.

"Deutschland?" He was fooling with a potato. He had a half-dozen small potatoes on his plate, still to go.

"*Was gibt es Neues?*"

"My brother is planning to go to England when his visa comes. Soon," Manfred recited. He had two letters from Kurt in his desk drawer and, stuffed in his wallet, a newly arrived photo of him wearing his white silk scarf.

"The sooner the better. There's not a minute to waste."

"Also, my father is changing work." That news had also come from Kurt. Lester grunted.

"If you ask me, I think your brother belongs here," Max said. "England is no place these days, either."

Another shrug. "It's a problem with the visa."

"We'll fix it."

"He wants to go to England. Cambridge is there."

"And your mother?" Lester asked.

"She's fine. She just had a hysterectomy." The word lay like a cold stone at the bottom of one of Kurt's letters, in a postscript.

"Oh?" Lester said.

"She's doing very well," Max put in. "Everything's in perfect order." He was not guessing. Julius Vogel had written to Max, asking him to share the news with Manfred, discreetly, without alarming him.

"Do you know what a hysterectomy is?" Lester asked Manfred.

"'Surgical removal of the uterus,'" Manfred quoted, word for word.

That's what it said in *Eugenics and Sex Relations*, sitting on Max's end table in the passion den, when Manfred sneaked a look; it wasn't in Cassell's, and Max, sharing the news with perhaps too much discretion, had not been specific.

"And uterus?"

"'The organ of female mammals in which the young develop before birth,'" Manfred continued. That wasn't in Cassell's, either.

They sat in silence for a few moments, Max looking a little embarrassed. "Well, I guess it's coming to all of us," Lester said. "I mean change of life, that time."

"It happens to all women," Manfred said without expression. "It's called menopause." He had memorized the entire entry in *Eugenics and Sex Relations*.

"Do you know how to spell the word?" Lester asked. "You should learn how to spell words like that. It's good exercise. It'll help you to improve your English." There was another silence. "I remember Stewie and Babs punting on the Cam when we all went to England," Lester then said. His eyes were a little bloodshot. "Who would believe it's Hitler's third anniversary soon?"

"I believe it," Manfred answered.

Lester took a long sip of tea, before changing the subject. "We're depending on you to come to the opera with us. We're saving a seat for you. Don't let us down."

"If you like," Manfred answered.

"We were in Bayreuth in twenty-seven. Not a good year at all."

"I never went to Bayreuth," Manfred said.

"In this town, Met tickets are like Bayreuth. Gold nuggets," Lester said. "I inherited ours. This year, Flagstad, Bori, and Ponselle. You'll use Stewie's ticket."

"I could start on a visa tomorrow with Judge Levy," Max said. "Just get your brother to say the word."

"If you were in Bayreuth," Manfred said, "you were not so far from us. I was six years old then, seven, something like that."

"Actually we stayed in Frankfurt overnight. I hated it. I loathed everything about the place."

"It's Frankfurt," Manfred said. "It's Hesse, it's Germany. Now tell me," he went on hurriedly, leaning forward on his elbows, swallowing his words. "Do you know how Charlie McCarthy got his circumcision?"

The brothers-in-law looked at him blankly.

"You don't know?"

They remained silent, waiting.

"I'll tell you," Manfred said. "He got his circumcision with a pencil sharpener. Get the point?"

"Oh, God," Max groaned, deadpan. "Another sharpie. Where'd you hear that one?"

"Adele told me."

"You have to repeat everything Adele says? Monkey sees?"

Lester laughed. "It could be worse," he remarked.

In the car later, while Lester lighted up a Cuban stogie, they listened to the news on the radio. It was the same as the morning news. Mussolini announced again that the Italians had broken loose in Abyssinia. Their airplanes—and Roman courage—had finally found the right targets. Haile Selassie was on his way to Geneva to give a speech to the League of Nations. Despite success, Mussolini was demanding additional sacrifices from the Italian people. That went for Italian people everywhere, including the USA. "The nerve of that bambino," Lester said, hissing through his teeth. In Germany, the Nurnberg Laws were now in full effect. No one was immune. The nation, according to Goebbels, would soon be One People. It was also disclosed that, on the other side of the world, swords were being raised in Japan, cries of banzai heard throughout the islands. Manchukuo was the target, China the victim, a sorry old story repeated endlessly, Lester and Max agreed. Meanwhile, six thousand homeless people were roaming the streets of Baltimore. Each night they slept in open doorways, in vacant lots, some of them in the gutters. The police had taken a head-count. And yesterday— to conclude the news—the Orioles had traded two players for an outfielder named George Puccinelli. No money was involved in the deal; the team had high hopes, as always. Some music came on. Lester switched it off.

Manfred, scrunched up in the tiny new-model space behind Lester and Max, could taste the sauerkraut that had come with the brisket. He was sitting on one haunch, tipped on a cold piece of metal. His leg cramped, a toe flexed on its own. Maybe he should eat more bananas, as Adele and Florence were always insisting. They forced bananas on him every day, for potassium. All his muscles ached now, as though he had been killing himself that afternoon at Hagenzack's with the rest of the slaves. Stuffed there in the luggage area, thinking of strange, permanent things like hysterectomies, Manfred began to make throaty animal sounds to himself. He beat his heels on the floor of the car, growling as they drove along Liberty Heights Avenue. Hyster-ec-to-my, he spelled to himself. U-ter-us. Men-o-pause. Subjects on which Dr. Schwabacher was an expert. Then: One Peo-ple. *Ein Volk.* He would like to have heard Kurt's comment on One People; Oskar's, too; rasped out to each other in the Vogel living room, in mocking voices, the restless voices of the family worldlings. Up front, Lester was haranguing Max about the erosion problem around Ashburton Lake. "The whole damn thing will cave in one day," he said. "You'll see, there'll be hell to pay." Manfred made a sound, beat his heels again. No one heard him. "That hill's been sliding away as long as I can remember," Max was saying. "A little more won't hurt. And you don't have to race the trolley. You know you can't win." The supercharged engines roared as Lester forced the car ahead. "Cut it out," Max cried. "You'll kill us all." It lasted for two blocks, then Lester slowed down, the trolley pulled ahead. "That's more like it," Max said, subsiding. A few minutes later, the car coasted down Fairfax Road at five miles an hour, while Lester puffed dreamily on his Cuban cigar. Alongside him, Max was half-asleep, his head skewed against the window. The car pulled up to the house without a sound. Manfred stretched, feeling his bones crack, feeling more tired than he had since his arrival in Baltimore. Nostalgia pulled at him, heavy as gravity, coming in like a tide, the unexpected image of his mother's secretive smile pulling at him, pulling him down. He was nearly suffocating from the power of it. A mere smile, a simple muscle twinge, a reflexive spasm of the psyche. "You intend to sit back there all night?" Max asked from the pavement. Manfred hadn't even heard the car door open. He scrambled out of the backseat, saying a hurried good night to Lester, thanking him for the movie and for dinner. His blood began to flow again as he moved up the path. "Come on," Max called. It was after ten, they were home, at least that's what they called it on Fairfax Road. Home.

Chapter Eight

"I might have known," Mr. Burns snorted, his eyes flashing green behind rimless specs. "A fuck book!" An acrid smile scabbed his face. He strode up and down in front of study hall, chewing his nails. Around him, a titter of disbelief rose in the air; then, as conviction took hold, it spread through the double room, front to back. He had actually said it. The assistant principal for boys had really spoken the word aloud. A fuck book! The boys couldn't help laughing, through waves of pleasurable shock. Sometimes they loved Mr. Burns and his special malice; it helped to make them feel alive. Outside, dirty Christmas snow still lay on the pavements. A single blue icicle hung from one of the study hall windows, dripping slowly. They had been back in school only a few days, after a two-week vacation. Energy was everywhere in the room, electrifying the atmosphere. They all felt it; it prickled their nerve-ends, set them pleasantly on edge. In the second row, Manfred Vogel stood facing Mr. Burns, his own face purple with mortification. A tiny grin appeared. It always seemed to appear at catastrophic moments, only to inflate the catastrophe. Through his mortification, Manfred observed it inwardly; at the same time saw Mr. Burns's obliterating face thrust at him like an axe blade; heard too the laughter and easy contempt around him. It was a majestic sin at Forest Park to be caught; it was unforgivable; you were shown no pity for it, but it could make you an immortal. Manfred didn't have to be taught that.

Manfred had finally reached the inner circle. Billy Brent let him borrow his beat-up little books for twenty-four hours at a time. Laura Piscitelli no longer heckled him. Her sarcasms had lost their edge. As far as Manfred could tell, he now seemed to be mysteriously beyond ruin in her eyes. He had proved himself by growing familiar to them all, by being steadily there, among them, day after day; and Laura and the rest had grown used to him at last. It was finally his right to be as dirty as anyone else. He carried the books, with Billy Brent's cool permission, rolled up in his pants pocket, along with his handkerchief. He read them at night in his long skinny room, next to Adele's. While she innocently popped her gum and combed out her hair after finishing her homework downstairs, he sat at his Indian desk in a deathly silence. The door was closed, a sign on the knob outside read Men At Work. He stared obsessively at each borrowed book with his face two inches from the page. The print was faded, the pictures dim. He had to concentrate on the text line by line, it was sometimes so obscure, and carefully trace each quavering picture, dot by dot, in order to reveal the unrevealable. Some of the books were comics, some were real life. Real life was more agitating, and stranger, too. Who were these indisputably live people, posing naked to the world? How had they come together, torsos locked? They looked so pained, eyes turned stiffly away from the camera, unsmiling and thin-lipped, as though they were terrified of being caught and punished for life.

Manfred's new eminence on Fairfax Road was reached during Christmas vacation. Life always changed then. The routine of school vanished, ordinary schedules were dropped. They were all at liberty together, barring certain obligations; and the time ahead seemed suddenly endless. A kind of sweet, unexpected languor set in, a sanctioned laziness that gave them all a chance to sit back and look at themselves every now and then without too much distraction. Manfred worked half-time at the store every day to help meet the rush of new business (it was the best Christmas ever, Max kept bragging, the new slogan was really bringing the customers in), but lunch was spent alone down at the harbor on Pratt Street, eating one of Florence's sandwiches under the shadow of a stubby black ship that had just arrived from Djakarta. The ship carried a load of spices. Each day the blunt, shining hull rose higher and higher at the dock as its cargo came over the side; a heady smell of cinnamon and saffron hung over the waterfront; and the busy crew, wearing navy pea jackets and wool hats, all spoke Dutch to each other as they worked to unload the ship. Tough consonants dropped like pebbles at Manfred's feet. He could understand some of the orders, shouted from above. He recognized old obscenities. For the moment, sitting there above the oily water, bundled up in his windbreaker on a freezing hawser, he might as well have been in Bremerhaven. Dreaming, he pretended that he was. He found himself stretching the lunch hour out by five or ten minutes each noon, sunning himself alongside the gleaming Dutch ship, eavesdropping on the crew, thinking of Germany and Frankfurt, of home, before finally overcoming his holiday inertia and slowly walking the few blocks back to M. Gordon & Son.

Almost everybody had something special to do over Christmas. Adele herself was passionately buried for part of the time in a copy of *Great Expectations* that she was reading for extra school credit. She kept at it for hours at a stretch some days, and by suppertime her eyes were so weary and distant that she was blind to everything else that was going on at the table. (Upstairs, her room held small piles of books that Manfred sometimes glimpsed through the open door; even poetry.) Joel Friedberg went off to Philadelphia for a two-day camp reunion; little Kitty Dean was learning how to crochet; and Billy Brent himself was working at Reads Drugstore to earn some extra money. Jojo McAllister, who spent the first couple of vacation days slumping up and down the street in an agony of boredom, was finally shipped out to Hagerstown by the McAllisters to visit his grandparents; his mother packed his German grammar inside his underwear, just in case. Jojo's misery was fair warning to everyone. Time could hang heavy if you weren't careful, liberty weighed a lot. It took a certain talent to be able to handle it.

During vacation, urged on by Florence and Max, Adele took Manfred to his first Baltimore parties. She let him tag along with her dates, sitting politely in the back of their cars, feeling like a little boy playing piggyback, worrying whether he had used enough mouthwash to keep his breath sweet for the whole evening. They went to a party almost every night; some nights they went to two. It was the thing to do during Christmas vacation. Mostly, the parties took place underground in newly built club basements. New club basements were everywhere that year. Each night a rented jukebox throbbed colorfully in a corner of another basement, while the needle slipped onto the latest 78's and everyone danced on the slippery congoleum. Adele lost her remote look then. Her energy returned. *Great Expectations* was behind her, Dickens forgotten. She seemed happy again, smiling confidently over her partner's shoulder in Manfred's direction. She usually had a new partner for every party, older boys, older than Billy Brent, one or two of them already in college: Sammy Billig, B.P. Potter, Len Schwartzman. Dancing, she liked to cock her forehead against her date's and tilt it a little to one side. She especially liked to dance like that with B.P. Potter, who seemed to be just the right height for her. It made Adele look alert and dreamy at the same time, typically, head slanted against B.P.'s or Sammy's or Len's, eyes a little opaque, unseeing, as though she was part of the world and outside it, as well; it was like one of the paradoxes she said she loved so much. The two of them rested like that on the dance-floor for a second or two, trying to find the right fit, then together, in each other's arms, they moved perfectly on top of the beat, Adele's left foot just slightly turned in, pigeon-toed. It gave her a nice, regular sway that everybody admired.

Manfred watched them from across the room, affronted. It was hard enough to sit in the back of the car and listen to Adele banter easily with her dates as they drove to and from the parties. He pretended to be urbane and knowing about their conversations, which he secretly condemned as empty chatter, laughing at all their jokes, making jokes himself from time to time. But it was far worse standing there with his back to the basement wall, alongside one girl or another, pretending to be impassive and poised as Adele and her date moved by him, wrapped smoothly together. He could never let them out of his sight then. Night after night, Adele danced past Manfred with her arm curled lightly around B.P.'s neck; they looked as though they had been at it for a lifetime. Where had they learned to be so easy with each other? How was it done? He was jealous of her and her new partners. He couldn't help himself. On his own, Manfred had discovered how hard it was to be a leader. That was what Adele had told him to be when they practiced together in the breakfast room to music on Florence's portable. Lead! she insisted. Grab hold of me! That's what girls want! Later, at the party, it turned out that American girls did not necessarily want what Adele Gordon said they wanted. Some pulled stubbornly away at Manfred's first touch; trying to hold them, he could feel them resisting in the small of their backs. Others bounced in his arms with notions of rhythm that they carried around in their heads rather than their bodies; they were always thinking about the music in a calculated way, clicking their teeth without knowing it or making little sounds on the roofs of their mouths to keep the rhythm. A few tried to lead *him*, their real nature triumphing over social convention; and Manfred followed or else. He worked hard at it, wanting to succeed; he was at attention all the time, obeying Adele's orders, happy to obey her orders, thinking out each step, faithfully planning every move for Adele's approval. The girls wavered at his touch; so did he. When the music began, they found themselves a step behind. Distracting new perfumes hovered around his partners' earlobes. Sweet rosewater, fading gardenia, rusty talc, a rainbow trove of cheap aromas, all wiped out soon enough by the reality of body sweat. Where was Adele? he thought in a panic, looking around the basement wildly.

Lead! he ordered himself. Do it your own way! One-two, right, he counted, grabbing hold, just as she had showed him, one-two, left, all managed to the smooth sound of satin trombones, then a nervous try perhaps at a watered-down Lindy, far too genteel, and finally, at the end, a sudden, swooning fall into a dip as the music died. That was when the whole point of the dance came clear. That was the moment when he understood everything. It was what they all looked forward to downstairs in the new club rooms, what they all seemed to need above every other thing, a brief, secretive pressure of penis against pudendum, hastily applied, never acknowledged aloud. It was also one of the specialties that Manfred was ardently practicing in the breakfast room with Adele's arms absently wrapped around him. She really knew how to dip, she was ardent, too, he thought. When he leaned over her, lingering perhaps a second too long, her hair smelled of soft baby shampoo, her back arched low in the crook of his arm, and her thighs unexpectedly opened to his while she tried to look away. But, holding her, he could read nothing in her face. Her mind seemed empty, her eyes without expression. She was not thinking about Manfred. I love you, he said silently, feeling no surprise. But of course there was no answer.

The parties they went to were tribal. There were no outsiders, no Brents, no Piscitellis, Deans, or McAllisters. It was the tradition, the ideal of an exclusive life expressed in unbreakable rules. Everybody from the northwest corner of town understood it. They came from the suburban homes of Forest Park, from the grander reaches of upper Park Heights, and the dark rowhouses off Lake Drive, across from the reservoir. The rest was off-limits, restricted by real-estate covenants; that included Roland Park, Guilford, Homeland, Loch Raven, Alameda, Catonsville, Dickeyville, Towson, Timonium, and Sparks; most of suburban Baltimore. But it was clear that, within the limits, Adele knew a lot of people, and that they all seemed to know each other. They had known each other most of their lives. So had their parents. Sometimes it went back generations. That was a part of the

Baltimore story that Manfred was learning, night after night, the tightness, the bonds, the way it all seemed to hold. At every holiday party, the city came closer together for him, individual linked to family, family to clan, clan to tribe. It was like connecting the dots in the Sunday funnies puzzle—something he loved to do on the sly—where the whole picture slowly came clear with each new stroke.

Manfred was a novelty to them all, he was the new German boy with the ruddy European cheeks and thin oval face, a little enhanced, even a little ennobled in their eyes by the strangeness of Europe and a history of dangerous unknown adventures. He had had a narrow escape, he was one of the lucky ones, his life had been saved; that was what they said among themselves; that was what their parents told them. Everyone flirted with him, trying to be normal, everyone teased, and Manfred began to like it, flirting and teasing back. A girl named Sally Krieger, who lived on fourteen acres out in the Valley, called him every day for a week until she was whisked away to Florida to celebrate the New Year with her parents. But first, kissing him on somebody's porch at midnight, she thrust her tongue deep inside his mouth and fluttered it there like a pink bird in flight. "Ja," Manfred heard himself say a moment later in an unrecognizable voice. "Yes, yes." "There's more where that came from," Sally Krieger said, with a matter-of-fact assurance that startled Manfred. Then she broke away after a second kiss to return to her date. After Sally left town, he found other "dollies"—as Adele named them, with what Manfred decided was just the right jealous note—who were willing to flutter their own tongues in mysteriously experienced ways. How confident they all seemed, how detached and calm, until he found the courage to reach out and touch them with the newly discovered power of his stroking fingers. That power was as amazing as everything else. It made the "dollies" tremble; it shook them. He added their names to his address book, alongside Bondi, Heller, and Marks, made calls to them himself. Nina, Louise, Ilene, the list grew. Like Florence he found that he could talk on the phone for an hour at a stretch, pulling out the sweet, masked conversations like taffy; and the "dollies" had just as much to say to him, full of double meanings and philosophical parries that passed for deep thinking. Max complained that he was wasting his time, there were better things to do than yakking on the phone with his hand cupped around the mouthpiece so the Gordons couldn't hear what he was saying; Adele laughed knowingly (and with a sharp possessive edge, he told himself, again and again); Florence herself looked amazed. The pink bird had nested in Manfred's brain; in his dreams the world dipped in unison to the music of Larry Clinton; the wet orifice of the world was suddenly everywhere.

Running from one party to another during the holidays became Manfred's whole purpose. Playing at night somehow consumed the entire day. It filled his consciousness, spilled over into sleep, where he could repeat the evening's adventures at his own slumbrous pace. Adele was his sophisticated duenna, self-appointed and full of generous authority, but she was erratic in her attentiveness and sometimes unreliable about simple facts. He had noticed that in her before, in other matters; it gave him an excuse to keep an even closer eye on her. Adele seemed to know everything, she could sum up a reputation in three unforgettable words, could predict what each party was going to be like before they got there. But sometimes, teaching him how to dip in the breakfast room, hiding her yawns while issuing urgent commands, a kind of irritable boredom seemed to come over Adele; lethargy soon set in, a powerful inertia; he recognized it in her as he had learned to recognize it in himself; she went slack in his arms, floated off with a now-familiar unseeing look, as good as blind, and snapped the radio off, claiming not to care one way or the other about parties. What was one more silly evening in the face of eternity, she cried. Who cares? It was shallow, that was what it was, it was inane. Who could possibly care? Then, leaving Manfred empty-handed, she would disappear upstairs to pick up *Great Expectations* or some other book where she had left off.

But once they were out for the night, once they had arrived at another strange house hung with balloons down in the basement, Manfred could see that the pink bird thrived for Adele, too. Anybody could tell that touching B.P. Potter, as they danced together, Adele Gordon was not indifferent to him or to parties. B.P. attracted her, parties were stimulating. Her restlessness vanished then for a couple of hours. Forehead slanted against B.P.'s, she became acquiescent for the rest of the evening, dancing in his arms to the jukebox music with a satisfied little smile for everyone. Manfred loved that smile, it seemed to speak of happiness and a kind of unself-conscious ease. It almost made him forget the other things, Sally Krieger and all the other dollies among them. But her happiness could turn, the radiance fade in an instant. It happened all the time. You could never tell with Adele Gordon. There were no assurances where she was concerned.

Every night during the holidays, out in the world with everybody else, at parties around town, they had the chance to try on other faces with no commitment; Manfred and his dollies, Adele and the new boys; as opposed to Fairfax Road, as opposed to home, where they always had to be themselves because at home you couldn't fool anybody, at home everybody knew.

About knowing: around this time, belly-whopping in the snow down Oakfield Avenue with Laura Piscitelli stretched out on top of him, Manfred heard Laura breathe the following words into his right ear just as they were making the grand turn into the woods that lay at the foot of Fairfax Road.

"I think Adele is trying to give Billy Brent the old heave-ho," Laura said.

The sled ploughed on, slowing against the drifts. It was very cold, Manfred's muffler was choking him, his galoshes were wet. It was their sixth straight downhill run together. "You hear me?" she asked, as the sled came to a halt.

They rolled off together, Manfred sitting in the snow for a moment, fooling with his scarf. Laura stood over him, waiting for an answer. She looked huge, silhouetted against the sky. "I know all that," Manfred finally said.

"You know what?"

"What you said."

"How do you know all that?" she asked.

"Adele told me."

"Buh-lawn-ya."

"Adele tells me everything."

"What did she tell you?"

"She said she wants to give Billy Brent the old heave-ho." He didn't even know what "heave-ho" meant.

"You're just repeating what I said."

Manfred got to his feet and began to brush off his corduroys. Just beyond the drift, in the woods, a silver-and-grey birch stood bare. Manfred could make out each piece of peeling bark, each brittle twig inked against the grey sky. There were no leaves anywhere.

"Well, Billy can't depend on her anymore," Laura went on, after a moment, sounding unusually thoughtful. "That much you can be sure of. I could see it coming a long way off."

"I don't want to talk about Adele."

"She said it's the end of the garage."

"So what?" As though he understood.

"You know about the garage?"

"I know everything."

"I watch. I watch from the alley. Kitty is the lookout."

"You watch what?"

Laura looked suddenly suspicious. "I thought you said you knew. I thought Adele told you everything." Laura waited a moment, then plunged on. "That's their place. There's an old rug in there from your mother's cellar. There's an old pillow and an old vase that Adele keeps paper flowers in. You tell Adele that I watch and I'll kill you. You see this?" She stuck her pinkie and forefinger in the air and waved them at Manfred. "Don't forget it. It's a terrible warning. Don't tell. Don't say anything." She paused. "Listen, you pull the sled up this time," she went on, in a suddenly bored voice. "I'm tired of pulling." She began to walk up the hill alongside Manfred. "On the other hand," she said, sounding thoughtful again, "Adele could always change her mind. She's always doing that. You know Adele. Hot. Cold. Some afternoon, you can watch too, if Adele and Billy . . ."

The snow was everywhere. It even seemed to be in Manfred's head, white and spinning. Already, he was trying to forget what Laura had told him. He just pushed it all aside, words, phrases, whole raspy sentences, everything.

The next day, Billy let Manfred keep *Flash Gordon* overnight. A few days later, Manfred had *The Gumps* to himself. Then he began to work his way through Billy's entire collection, as he pleased. "You sure you understand all that stuff?" Billy asked. "I don't want to be the guy to start you down the primrose path." Billy smiled a tiny smile, self-conscious and tight. They were standing in the middle of Fairfax Road, slush around their feet. Billy had the afternoon off from his job at Reads. He was holding his football in his hand, even though the game had quietly given way over the holidays. It was too late in the season, the weather was impossible for play, nobody had the time. Billy spiraled the ball in the air, Manfred reached for it. "Better fold that book up a little tighter," Billy advised, pointing at Manfred's pocket. "Or you're gonna drop it in a puddle." He tossed the ball to Manfred. Manfred tossed it back. Then Manfred tucked the book out of sight. "One more," he called as Billy hesitated with the ball. Their feet were wet, it was late afternoon. Ignoring Manfred, Billy slouched over to the curb and began to trace a letter on a car window with his forefinger. A huge *A* appeared. "Where's your sister?" he asked as Manfred approached.

"I don't have a sister."

"You know what I mean, Adele."

"Adele's not my sister."

"What is she then? I thought you were related." He drew a small *d* next to the *A*.

"I don't know where she is. We're not anything."

"Go get her."

"No," Manfred said after a moment's hesitation.

"Come on, we'll toss a few around." Billy spun the ball in the air again, temptingly.

"If you want her, you go get her."

"No sir. Not me."

Manfred was silent.

"I've never been inside that house and I'm not starting now," Billy said.

"What are you talking about?"

"Don't play so dumb. Your mother and father don't want me around. They wouldn't let me through the front door. I'm all the wrong things. Not Jewish and not rich."

"Who told you that?"

"Nobody had to tell me. I *know*."

"Just ring the doorbell and ask for Adele."

"Not me."

Manfred hesitated. "Well, I've never been in your house either," he said.

"Nobody has. It's the black hole of Calcutta in my house. My old man won't turn on any lights. We eat supper under a twenty-watt bulb. Now where is she?"

"She's probably fooling with her hair or something."

"Party time," Billy said.

"We're going out to Old Court Road tonight."

"Rich kids." Billy then made an *e* on the car window. "Go get her."

"No."

"Come on, Fritzi."

"My name's not Fritzi, Willy."

That brought Billy Brent to. His head snapped up. "If you had an American name," he said defensively, "I probably wouldn't have any trouble. Tommy or Bobby or something like that."

"My name's Manfred." Manfred then added an *l* on his own to the car window.

"I know how to spell," Billy said.

"A-d-e-l-e," Manfred said, finishing it off.

"Who asked for your two cents?"

Billy bounced the football off a tree trunk while Manfred added his own name to the window, M-a-n-f-r-e-d, in sloppy letters. Then underneath he quickly wrote in B-i-l-l-y, connecting the three names with a wavering arrow that immediately began to run. "Go get her, we'll toss around a couple," Billy said. Manfred didn't answer. They began to shuffle around again, unwilling to leave the street. Every now and then Billy glanced up at the Gordons' house, towards the passion den. The red brick on the front of the house was faded, like all its neighbors, the porch trim needed a coat of paint, windowpanes shimmered opaquely in the cold light. Behind them now, Hilary Kates began to practice the piano. First she did scales, then arpeggios, slowly. A few houses down, the widow Junkermann peered out at them through her living room window. She wore a wrapper from Japan; she always wore a wrapper, naked underneath. "You guys gonna play or not?" somebody yelled from up the street. It was Laura Piscitelli, striding towards them. She was wearing her earmuffs. Her father trotted behind her, squinting angrily out of his nearsighted eyes; behind them, a few yards farther back, was Mrs. Piscitelli, wearing her new sheared-beaver coat and a fur hat. They were heading home.

"Well?" Laura called in a hoarse voice. "Well, what?" Billy called back. "Don't just stand there," she ordered. "Do something." She was laughing. "More lousy jokes," Billy muttered, turning his back on her as she came closer. "Let's have some action," she yelled. "I want to see some action." "She really gets on my nerves," Billy said. "This whole street gets on my nerves."

Manfred watched the Piscitellis pass by, in a row, one a couple of yards behind the other, as though they were linked together by a chain of improbable accidents. Mr. Piscitelli was carrying a bag of groceries. He kept shifting it from one arm to the other. The lower half of his face was squeezed together like a toad's; he hardly had a chin; his two curving nostrils were like black tunnels carved into his head. It was Italian fate to look like that, the devil's punishment; everyone held it against him, as though he had made the choice himself. Even his daughter made sarcastic jokes about it. She considered him personally responsible for looking like a toad. Laura was already taller than her father. So was her mother. Her mother, stately and aloof, was able to peer down at the top of her husband's head. It seemed unfitting for a wife to stand taller than her husband, it suggested something unnatural. Mr. and Mrs. Piscitelli looked to Manfred like one of the horribly mismatched couples who clung together so guiltily in Billy's real-life books. They were just as compelling, just as awful.

The afternoon closed in, the light washed away. It was grey everywhere. Down the street, the Piscitellis' front door slammed shut as they entered the house. The street was quiet, except for the sound of *Für Elise*. Soon the holidays would be over. "It's dark before you can even say boo," Billy Brent said. "I hate winter."

"Yeah," Manfred said, slowly turning towards the Gordons' porch. It seemed to take forever, just to make a right turn away from Billy Brent. "I guess I'd better go in and get ready for supper," he added.

Billy sauntered away. He jumped up suddenly, football tucked under his arm, and touched the top of a lamppost with his free hand. Across the street the names on the car window were blurred, the arrow had disappeared, it was all unreadable. On his way to the porch, Manfred cupped his hands over his mouth and made a strange sound. Billy stopped in his tracks. "Where'd you get that?" he called. Manfred didn't answer. "Hey, I'm talking to you," Billy snapped. "I just made it up," Manfred said, climbing the porch steps. He was smiling to himself. From the porch he could see the widow Junkermann peering out from behind her curtains. Next door, Hilary Kates banged a chord full of wrong notes. Manfred had made the strange loon sound just right; he had practiced enough in the bathroom. It made him feel clever, clever and a little superior. "Another conceited snob," Billy called out bitterly from the pavement. "I knew it from the word go. Everybody on the block knows it. Rich kids. And that goes for all of you, the whole family, your sister included."

As though Manfred had to be told.

Chapter Nine

A fuck book.

Mr. Burns stalked Manfred. He circled him first, eyeing him up and down, flirtatiously. He paused a moment behind him, mimicking his posture, getting another laugh from the boys. He strode down the aisle to the front of the study hall, folded his arms, pushed out his chin, like Il Duce reviewing his legions. He shook his head in awe and wonder.

"Hot damn," somebody said, hyperventilating in the back of the room. Nearby a book fell on the floor. "Hold on," Jojo whispered to Manfred from the desk across the aisle. His German grammar lay upside down in front of him. "Don't let him get to you."

"Something on your mind, young man?" Mr. Burns asked, rising on his toes and looking pointedly at Jojo. "Something we should all share, perchance? You there, shorty, what's your name, I'm talking to you."

"I didn't say anything," Jojo said.

"You didn't say anything," Mr. Burns said in a wounded voice. "Oh, my. Butter wouldn't melt. Seems to me I clearly heard the sound of a pipsqueak. Shapiro, from your present vantage point in life, did you happen to hear what's-his-name shorty speak just now?"

Unfortunately, Shapiro's present vantage point was the seat right in front of Jojo. It was impossible to have missed a word. He nodded an assent and looked miserable.

"And you, Mr. Halliday?" A little more politely here, even ingratiating, as though the question were being asked of an equal.

Mr. Halliday—Forester, first vice-president of the Boys' Leaders Club, justice of the school court, champion trackman, and preparing, next year, to enter Colgate—sat on the other side of Jojo. Mr. Halliday, of course, had heard; and so had Rosen, Offit, Corkran, Miles, Steele, and Klepfish, as each reported, without enthusiasm, after being asked the same question. But the affair had been going on for too long. Even in Mr. Burns's terms, which always made room for an extended display of pleasure in the other man's agony, it had been going on for too long. Jojo was a diversionary tactic, hardly worth an assault. They all knew that. Another book dropped. A desk top banged shut over on the left. Up front, Manfred shifted from one foot to the other, weary of standing. One toe had already cramped; he wanted to scratch his eczema; and the seat of his pants was stuck in the cleft between his buttocks. Mr. Burns clutched the horrible little book as though it was a primitive club, something a caveman out of prehistory might have carried. It had dropped out of Manfred's pocket twenty minutes before, as he had reached for his handkerchief, and Mr. Burns had picked it up with an expression of open joy. Ah-*hah*, he had said several times, turning the pages slowly and pausing at each picture. It was one of the real-life books. "Ah-*hah*."

Manfred's groin burned. He desperately had to pee. His physical self was in full turmoil, flaming with public embarrassment. His face went from red to purple to red again. He reached behind, tugged his pants loose. Something in his eyes glazed over.

Mr. Burns turned back to Manfred. "You're not even a regular student here, are you, sir?" he asked.

There was no answer.

"You there. You, Heinie boy."

"Yes."

"Yes what?"

"Yes, I am not a regular student."

"You are a guest, then, of Forest Park High, a guest, one might say, of all your fellow students in this room?" Mr. Burns made a sweeping gesture with his hand, taking them all in, making them conspirators without their consent, then rested his hand on his hip.

Again, no answer.

"You are a guest then of Forest Park High?" Mr. Burns repeated, swaying forward as his voice began to spiral upwards. "You are, or you aren't?"

"Yes."

"And you have fouled the hospitality that has made a place for you?"

Manfred's eyes widened, his cheeks flushed again. You really had to watch out, there were lunatic Americans everywhere, lying in wait; reality in the USA was totally unpredictable, changing shape at any provocation.

Mr. Burns sighed. He shook his head, he chewed a nail, spit it into a wastebasket. Then he slapped Billy's little book hard in the palm of his hand. The cover came off. "May I ask you a simple question?" He stared around the room patiently. "That is," Mr. Burns went on, "if you have no objections?"

Manfred nodded blandly, shifted his weight.

"You do or you don't have any objections?"

Manfred nodded again. His little toe curled. Next to him, at waist level, Jojo seemed to be hissing under his breath.

"Tell me first where young gentlemen get their hands on literature like this." Mr. Burns came down heavy on the word "literature" and slapped the book again. Another page tore. "Then tell me second," he continued, "what young gentlemen do with themselves after they read it."

A rustle spread through the room. A kind of buzz rose in the air. A few boys began to cough. Mr. Burns had suddenly passed beyond malice with a single question, had leaped over the top into a no-man's-land littered with bomb craters and rusty barbed wire. No one wanted to follow; it might put their lives in danger. Mr. Burns stared at Manfred, Manfred stared back. "You *are* a young gentleman now, aren't you?" Mr. Burns asked.

"Yes, I am a young gentleman," Manfred said; but he said it in German.

There was another faint rustle, this one like a communal breath passing the length of the room.

"Hmm," Mr. Burns murmured thoughtfully. His mouth pulled to one side. His fingers fanned across his face. He waited a moment. "Another unhappy example of homo sapiens smart ass," he said in a muffled voice. "The question is what will we do with you. You're a weird one, all right. Tell you what, why don't you just say what you just said again. In the original. I dare you."

There was hardly a pause. *"Ich bin ein junger Deutscher Herr,"* Manfred said, enunciating slowly and expanding a bit on his opening remark. As he spoke, the shock of hearing the statement aloud washed over his face. Everyone in study hall looked stunned. The strange, foreign voice they heard was Manfred's own, speaking in familiar tones, but the words, the simple, flat, declarative statement that few of them understood, had come from Germany, across the Atlantic, from Frankfurt-am-Main. Maybe he was a weird one. Manfred stood there now, pulling in his stomach, thinking in German. He no longer cared about Mr. Burns. He no longer had to pee. His physical self was leashed. It was too late, in any case, for Mr. Burns. He had structured the hazing all wrong, his timing was drastically off. The boys in study hall were impatient. Mr. Burns's victim had mysteriously evaporated. It sometimes happened that way in the face of unexpected resistance. The four-minute bell rang. Feet scraped the floor. There was a sharp sound of books being piled together. Shapiro sank back in his seat in relief. Halliday smiled grimly. The rest of them pretended to be distracted. Mr. Burns understood. He stood there rigid, becoming his own victim, copper face darkening. He was not used to losing a quarry.

"Quiet out there," he shouted. The room stilled. "You will move when I tell you to move and not a minute sooner. You are here under my jurisdiction and my jurisdiction in this room is the law. Any of you disagree with that we can have a little chat downstairs in my office. You got that, Corkran?" Corkran had bent over to tie his shoelaces. "Sit up at your desk, sir. All of you, sit up at your desks. It's disgusting to even look at you. Round shoulders and pot bellies. Flab. Slackness. Nellies, to a man." He began to pace up and down in front of the room, fixing them one by one out of the corner of his reddened eye. "And as for you," he said, stopping in front of Manfred again, "whatever your name is, Vogel, you are on probation. P-r-o-b-a-t-i-o-n. A serious matter in these American parts, you should know, designed for worst cases only. And as for the rest of you horny bastards"—there was a muttered cry of protest from some of the boys at this—"as for the rest of you self-described gentlemen, you will NOT carry such trash in the corridors of this school at any time, for any reason, you will NOT be caught reading it in study hall, in class, on the toilet, or anywhere else, you will NOT bring your foul smut into my presence or anyone else's presence, you will NOT . . ." The final bell rang. Manfred's knees sagged. "Probation, my friend," Mr. Burns called, waving the fuck book at him. "And un-der-pe-nal-ty-of-ex-pul-sion."

Jojo stood up. The whole class stood up. "Don't say *anything*," Jojo said, through his teeth. "Piss on him."

It was over. The boys began to move around Manfred in sickly waves. There was a rush to get out of the room. Mr. Burns put Billy Brent's book in his pocket and backed away. Mr. Halliday had a question for him about the agenda for the next meeting of the Boys' Leaders Club. They began to confer together at the door to the classroom. Blackboard images hung in the air, the smell of chalk and erasers was everywhere. A geometry theorem was written out on one board in a clean hand; beneath it, some arcane Latin declensions. On the right wall, a map of the USA, all pink and green and yellow, was pulled down. A map of the whole world, flattened out, hung next to it. It was all so regular, it was hard to imagine anything disturbing its ordinariness. Manfred clutched his books, walked out of the room with Jojo, past Mr. Burns and Mr. Halliday, who were standing face-to-face, at attention. Mr. Halliday smelled of Lifebuoy soap. Rivulets of sweat ran down each side of Manfred's rib cage. Jojo was jostling him. "Quit shoving," Manfred said, moving into the hall with everyone else. Jojo stuck to his side. Probation, Manfred thought, tasting the unknown word. Under penalty. Of expulsion. It sounded important.

"And what do you call yourself?" Miss Tubman asked later in the day, leaning against her grand piano, arms folded judgmentally.

"Manfred Vogel."

"I know what your name is. I'm talking about your voice."

"Baritone." It was a wistful shot in the dark, sounding manly, desirable. Alongside him now, holding a piece of sheet music, Billy Brent stood shuffling his feet. He was on edge, yawning with tension; it was Billy who had introduced Manfred to Miss Tubman and Boys' Chorus, fulfilling his promise, as he reminded Manfred a half-dozen times a day. Also, Billy had had word of Mr. Burns and Manfred from Manfred himself five minutes after the study hall episode had taken place. "You don't have to worry," Manfred said. "I didn't say a word about whose book it was." "You better not," Billy said, making an ugly fist at him.

"You're no baritone," Miss Tubman said to Manfred in the music room. "Why-do-you-all-think-you-have-to-be-baritones?"

"He's really a tenor like me," Billy offered nervously, his bad temper moderating for the moment.

"A tenor like you," Miss Tubman said, as though it were the name of a dread disease.

Miss Tubman seated herself at the piano. Her hair was purple. Her dress was purple. Purple was the chosen color of her life. A single rope of fake pearls hung to her waist, rising over a Himalayan bosom like a mountain snake. Two large rouge marks spotted her cheeks. The rest of her face was dead white. Miss Tubman also had an enormous curving swayback, tightly corseted for the classroom. Singing together while she stood in front of them, beating time, her students were sometimes unable to keep their eyes off Miss Tubman's swayback.

Miss Tubman played a scale, arpeggios, heavy chords. Her touch was not delicate. She thumped grandly, all rhetoric. "Try it," she said. Manfred opened his mouth, hoping to find the pitch. Do, re, mi, fa . . . do, ti, la, so . . . A thin, reedy sound filled his ears. It was the piercing line of his own voice. Perhaps he would be chosen as a soloist, it sounded so fine. "Do you read music?" Miss Tubman asked. "Yes." "Do you play an instrument?" "The piano a little." "What is that accent?" "German."

It was after school, a little late for strenuous effort. Miss Tubman had spent the day teaching "When the Foeman Bares His Steel" to five consecutive classes, to be sung in the original harmonies at the next Senior Farewell Assembly. "Mr. Brent may be able to help you," she said from the piano. "I mean," she went on, "that you-can-stand-next-to-him-for-the-time-being-for-better-or-worse." In the back of the room a few boys sat waiting for them to finish. "In these precincts," she went on, "we are serious about music. Are *you* serious about music?"

"I love music," Manfred assured her piously.

She called everyone together then to begin to rehearse "Smoke Gets in Your Eyes." Manfred stood next to Billy, sharing the score. Miss Tubman pounded the piano, declaimed the lyrics, sometimes looked stricken. "Oh, no," she cried out. "That will never do. It would break Mr. Kern's heart. Try again."

The boys sang out. The windows in the classroom actually rattled. "They asked me how I knew, my true love was true . . ."

He found out what probation really meant the following night, just before dinner. He had come home late from school, after another Boys' Chorus rehearsal. "They asked me how I knew . . ." Billy Brent sang, walking down Oakfield Avenue at Manfred's side. On Fairfax Road they parted, Billy turning up the path to his twenty-watt house.

As soon as Manfred opened the Gordons' door, before he even stepped over the sill, he could feel a faint sense of disturbance, the merest displacement, as if something ordinary in the Gordon family had been temporarily dislodged from its usual place in the house. It hovered there lightly as he hung up his coat, unidentifiable, and intensified when he walked into the living room. Where was his friend Max? He was not sitting in his corner chair, reading the newspaper backwards or reciting aloud interesting headlines that caught his attention. Adele crossed his path on her way from the back of the house. She was tiptoeing exaggeratedly, and when she looked at him she rolled her eyes—a terrible habit, Manfred thought, it made her look like she was going to have some kind of fit—and wagged her head in the direction of the kitchen. Manfred knew a warning when he saw one. He heard a bustle back there, a more-than-normal rattling of pots and pans. That confirmed it. "What's going on?" he asked. Adele rolled her eyes again and got out of the way. "Stay," Manfred said. "Nothing doing," she answered, heading up the stairs. Manfred walked into the breakfast room. Through the door, he could see Florence standing bent over the kitchen sink. She had stopped banging the pots and pans. The hem of her housedress hung crooked, her thin, bony legs poked out below. A smell of tomato sauce came from the stove. Unsliced white bread sat on the breakfast room table, a half-finished glass of milk stood alongside it. Manfred heard the door to Adele's room slam shut overhead. Then Florence lifted a pot off the sink and banged it feebly on the drainage board. It was like a signal for them to begin; and they did begin.

Max burst in from the dining room. He held one hand behind his back. His beautiful white hair was dishevelled, as if he had been running his hand through it, but it still looked beautiful. "So you're home at last, running around the streets night and day," he shouted. Florence lifted her head at the sound of his voice and peered timidly around. Her mouth curved downward, her eyes were wary. Timidity was not her way. What was happening? "And what's the meaning of this, if I may be so bold?" Max demanded. He was waving a piece of notepaper in the air. He shook it in Manfred's face. There was cigarette ash on his tie. "Don't give me that insolent look," Max went on wildly. "It's insulting. You're as much a part of this family as anyone else. You take, you have to give, too, What-is-this?"

"What-is-what?" Manfred asked stupidly.

"What's wrong with you, anyway, are you blind or something?" Max waved the notepaper at him again. Florence joined them in the breakfast room, standing behind Max, her hands wrapped in her apron where they struggled in agony with each other. For once, she seemed speechless.

"I don't know what that is." Manfred said. "What is it?"

"You tell me."

"How can I tell you when I don't. . . ."

"Don't be insolent," Max interrupted. "I won't stand for insolence."

"I am not insolent."

"And don't contradict me," Max said. "You don't know what this is, it happens to be a letter. It's a letter from your school. Now do you know what it is?"

"No."

"You're on probation for unfitting behavior. That's what it says, right here." He flicked the notepaper hard with a fingernail. "You're on probation under penalty of expulsion."

The fuck book: it would never be over, it would follow him to the grave.

"Well?" Max yelled.

"Max, please," Florence finally said. She looked as if she had one of her migraines. Three vertical lines between her eyes creased in pain.

"What in God's name have you done?" Max asked Manfred.

"Give him a chance now," Florence said.

"What is unfitting behavior anyway?" Max demanded. "What does it mean?"

"I was rude to my teacher," Manfred answered. "He was mispronouncing a German word. I corrected him."

"You what?"

"I corrected him in front of the class. Mr. Lambrino thinks he speaks perfect German. He's pretty good, but he's not that good. He's always picking on Jojo McAllister. He picks on me. He says we cheat together. He says I keep giving Jojo the answers. It's not true. I don't keep giving him the answers. He kept mispronouncing a word and I corrected him."

"Corrected who, for God's sake?"

"Mr. Lambrino." Manfred began to glow then with self-righteousness; already he partially believed his own lie.

"You think you're smarter than your teachers?"

"Max," Florence said again. She tried to smile at both of them, a patchwork grin made up of hope and craftiness and visible nervousness.

"Do you?" Max said.

"Do I what?"

"Are we speaking the same language?" Max asked, sounding as though he was ready to weep. He lowered his voice then, his energy ebbing. The first storm apparently was over, but his voice continued to tremble. "Do you have any idea how you've humiliated us? The position you've put us in? Condescended to by the assistant principal for boys. Threatened, even worse." His voice began to rise again. "Is that what you did in Germany, go around correcting teachers?"

"I don't think that's so bad."

"You don't think that's so bad."

"No."

"It's bad enough for probation."

"Then I'm sorry."

"Sorry?"

"Go easy, Max," Florence said.

"I said I'm sorry," Manfred said, full of virtue. "Anyway, nothing will happen. Half the school's on probation."

"What do I care about half the school? You're the issue here. They'll kick you out anytime they want."

"They won't kick me out. I'm already in the Boys' Chorus. That's an honor, ask Billy Brent. I coach German. Miss Feinstein asked me to write a special paper on the Versailles Treaty. I do all the blackboard work in French for Miss Klinesmith. They want me to give a speech to the dumb Deutsche Verein. They won't kick me out. What do you think, they want me to transfer to some other school, City or Poly?"

Max began to quiet down again. There was no arguing Manfred's point. He was doing good work in school, everybody knew it. Besides, passion had worn Max out; he was not used to dealing with it. His anger was not natural. It falsified him and he knew it. He was the born conciliator, but he could be stubborn. What rankled now was Athol Burns's challenge to Max Gordon's moral authority; and there was also Manfred's callow complicity in the whole matter, his matter-of-fact explanations, offered up as though Max were some kind of sappy idealist, as if he had no eye for reality. That hurt, that was a wound. The sense of disturbance stayed in the air. Max's moral authority saw to that as Max and Manfred wound down the exchange, each trying to get a few last words in; and it remained there all through dinner, as Florence tried to act as though nothing had happened and Adele shrewdly turned the discussion to the question of higher education. As subjects for debate, higher education, along with Zionism, anti-Semitism, President Roosevelt, and money, never failed at the Gordon dinner table; they made the digestive juices run, sometimes with curdling effect. Now that business was picking up, Max was saying in a slow, controlled voice, due of course to the success of the new advertising slogan he had invented, shouldn't they be considering alternatives to College Park, down Route 40, for Adele? Max's normal color had returned. He was holding himself in check. If she had a choice, he asked lightly, any school in the country, no holds barred, where would she want to go?

"Reed," Adele promptly said. "Or Sarah Lawrence. Listen, is this a serious question?"

"It could be serious," Max said. "I may have the senior-class ring franchise next year, at least for two schools. It just might be. I do push-ups at Hagenzack's with somebody from the Board of Education. He's hinting around, and so am I. It's beginning to feel right, it's not like a fix, maybe I'll get a contract."

"Well, if you're really being serious," Adele said, "the place I'd really like to go is Sarah Lawrence."

"You mean you'll be gone next year?" Manfred finally asked.

"What did you think we were talking about?" Adele said.

Manfred looked at her across the table. She was sucking on a strand of spaghetti, head bent over her plate. A drop of watery tomato sauce had settled on her chin. Sitting there, not making a sound, Adele was all fleshy circles, freestyle arcs, incurving folds. One tucked into another, rounded and full, there was not an angle or a straight line in sight. Her mouth, as it pulled the spaghetti in, made a perfect cherubic O. Manfred looked away. He could not bear to be caught staring at her. She could chill him with an offhand glance; she could put him where she thought he belonged, outside everything. And with the future now in sight, she was changing, that was clear. There were little pieces of evidence everywhere. Adele now kept a blue-and-white box of Kotex in open view on a bathroom shelf. The week before she had thrown away all her bubble gum, stunning Florence Gordon. She did her homework these nights in her own room. There was nothing Manfred could rely on. "I guess I never thought anything about it," he said. "It never occurred to me."

"Well, figure it out. I'm a senior."

"All right," Max said. "It'll either be a new house, or it'll be one of those schools, or it may even be something else. Can you get into Sarah Whosis?"

"You guys go buy a new house," Adele said. "College Park is fine with me."

"So you won't be here next year," Manfred said, watching Adele pick at her salad.

"If I'm at College Park, I'll be home every weekend."

"Is Billy going to College Park?" Manfred asked.

"Billy who?" Florence said.

"Billy Brent."

"Billy Brent is trying for a scholarship to Franklin-Marshall, Western Maryland, or Tidewater."

"Tidewater," Manfred said.

"Backwater would be more like it," Max said.

"How can Billy Brent afford to go to Franklin-Marshall or one of those places?" Florence asked. "His father's a conductor on the number 5 trolley."

"I just said he's trying for a scholarship."

"I didn't know Billy Brent was smart enough for a scholarship," Max said.

"He's not," Adele answered. "He's going to have to do it by being all-around."

"Well, in my experience, people on probation don't usually get to go to college," Max said, looking meaningfully at Manfred.

"Oh, Max," Florence protested. "We're talking about Billy Brent."

"What *was* that German word anyway?" Max asked.

"What German word?"

"I think it's time to tell them," Florence said to her husband. She cleared her throat. "And if you don't, I will," she added.

"What?" Adele asked.

There was a pause. "Your mother's pregnant," Max finally said.

Adele looked from her mother to her father and back again. Then she looked at Manfred. They stared at each other, their eyes feeding a slow mutual comprehension across the table.

"Your mother's going to have a baby," Max said.

Adele giggled. "Mummy." The old endearment, which had been put away years ago, escaped by itself.

"That's right," Florence said. She began to brush breadcrumbs into the palm of her hand.

"I thought you were too old," Adele said.

"Some people would say so. I'm forty-one my next birthday, you know that."

"A baby."

"That's what they're called."

Now Manfred began to blush. Odd shapes of red mottled his cheeks, just like Adele's. They both looked a little feverish.

"So you see," Max said grandly, "you have to go to one of those schools, we need the room."

"Oh, Daddy."

"There's plenty of time," Florence said. "Another seven months."

"You're going to have a baby," Adele said.

"The whole world has babies," Max stated.

"I just never thought of it," Adele said.

"Well, that's the way it's going to be and about time, too," Max said, smiling the most uncontrollable smile Manfred had ever seen.

"There's some nice Jell-O in the Frigidaire," Florence said. "Want some?"

"You stay where you are," Adele offered. "Don't you move."

"Manfred?" Florence asked.

"I have homework to do," he heard himself say. "I have to start on the Versailles Treaty for Miss Feinstein."

"Then hop to," Max said imperiously. "What are you waiting for?"

He jumped to his feet, eager to get away from the table. There was no reason to stay, in any case. He had just decided, within the last five minutes, that he would turn his back on all the Gordons, on Florence, and Adele (even Adele, especially Adele). He would pay no attention to Fairfax Road or Forest Park High or M. Gordon & Son or the rest of Baltimore, Maryland. Like Billy Brent, he knew when he wasn't wanted. Everybody in America knew how to live without him, he would learn how to live without them.

"Dear Kurt," Manfred wrote later that night at his Indian desk upstairs. "Strange things are happening . . . " He gazed at the wall in front of him and chewed on his pen. He traced the crude swastikas on the desk with his forefinger. His stomach rumbled.

"Already it is 1936. Where is 1935? It is all the same thing, yes? One year and another, like a spool of momma's black thread that goes on and on. Perhaps it would be good to finish unwinding the spool once and for all. I believe that sometimes. Everyone believes that sometimes. Still, you must not concern yourself about me. I remain, as always, cheerful. I am not Hamlet or Werther, either. I do not have the will to be crazy. Or the courage. I hold onto the future. Is that not what we believe in? The thread will not break, whatever the strain. I know because positive signs are everywhere. Here, for example, the trend is definitely upward. It is so, new times are coming. Local trading is strong. There are many pluses in the bond market, wheat too moves up. Stocks show exceptional vigor on the curb exchange. We are on the verge. So the newspaper says.

"Even so, some forces are centrifugal, as always. You have to be on the alert each minute. In Catonsville, a grenade from the war exploded in a young man's face when he picked it up, taking off his nose, and downtown, another young man jumped to his death from the roof of an office building, shouting of love. Of love, of all things! All over the country, meanwhile, the divorce rate is up, climbing to new highs, like the bond market and the curb exchange. Things fly apart. It says so in the newspaper, and Lowell Thomas confirms it. And they claim here that marriages are made in heaven.

"Already in 1936 the mother of the Italian girl Laura has given her gold wedding ring and two other gold rings to Il Duce to make ammunition against the Ethiopians. She gave them at a rally for the Imperial Legions that took place downtown in the Italian Gardens at 806 St. Paul Street. Mrs. Piscitelli kept one diamond—also from her wedding—for herself. A priest from the Italian neighborhood made an appeal in the Italian language. 'The day,' he shouted, 'when England fires her first gun at Italy will be the end of England.' Then all the women cried 'Vive Mussolini!', stripped the rings from their fingers—gold, silver, other things—and dropped them into a barrel to be shipped to Rome. I learned this in the newspaper. Ten pounds of metal were collected when the priest made the sign of the cross over the crowd and began to weep in front of everybody. In Paris, the newspaper also says, women wear clothes like the natives of Africa, with colored turbans and tall bird feathers on their heads. It is the new style at the Ritz and other places, to show sympathy for the Ethiopians.

"I have learned how to dance. (I warned you, strange things.) We dance in Baltimore on Saturday night, fast and slow. It is a serious matter. After we dance, Sally Krieger puts her tongue inside my mouth. That is the rule. My prick, faithful to its nature, grows hard. We push against each other on the back porch, in the dark. She is very strong. Her tongue is like a slippery eel. Oh, my God, she says, oh, yes, and I say the same thing, in order to sound enthusiastic. We are like two cats lapping up milk. When the dish is empty, we go downstairs to the basement to dance again.

"Sometimes I see the ships come in to port down at the Chesapeake Bay, from Germany and all over the world. They sail up from the sea, pouring smoke, a hundred miles to Baltimore. On one side is Maryland, on the other Virginia. There are twenty-three (23) counties in Maryland. Did you know that Baltimore is placed the furthest west of all the great seaboard cities of America? It is the most important fact about the city. It brings the ships close to America. There was even a ship here from Russia, the *Pushkin*, on Light Street. I saw it last week, with a red star on the prow. There was another ship from Hamburg, smelling of Germany, and in the spring the *Emden* will visit Baltimore on a goodwill cruise, so the newspaper says.

"Here is a recipe for Benno to try. It is from the South. He will like it, I think. The cook's name is Aunt Priscilla and she writes in the newspaper on the society page. Aunt Priscilla says, 'Try dese bread fitters fo' a change, Miss Reader, if you likes da sherry flabor in yo' cookin'. You shud hab bread dat's a bit stale fo' dese. Too fresh bread am ap' to be kinda soggy. Cut off de crusses an' on each side pore a spoon ob sherry . . it's mo' den pal'able fo' use jes so. Den fry.' Tell Benno I can send more.

"Soon they will make me leave school. That is the news here. The assistant headmaster is my enemy. He persecutes me. As a result, here at 'home,' Mr. Gordon, who was formerly my friend, now hates me. Now also his wife is pregnant. It is actually true. There will be a new baby mammal born in August. That is what they say. It is my fault, I think. They cannot want a new baby mammal at their age. The lady of the house is forty years old. Is that not time for menopause? Already she can hardly stand up, and she has little blue veins on her legs, behind the thighs, like a spider's web. They make me eat spaghetti and meatballs with them for supper, as in Rome. When the lady of the house makes the beds, a cigarette sticks in her mouth and she squints her eyes from the smoke. The same in the kitchen. We have ash in the soup sometimes. Every week, late in the afternoon, the girl Adele goes in the garage next door with her friend Billy Brent. Billy Brent lives down the street in a dark house, like a mole. He is an American giant with curly hair, six feet tall, as tall as Manitou the Indian. His father drives a tram. They think their secret inside the garage is still their secret, but I know because Laura the Italian girl told me. Sometimes Laura watches through a dirty window from the alley, and she tells me what she sees when we belly-whop in the snow together. On the sled, her arms pin me down from above and she breathes in my ear like *die Alte's* little cur, Pforzel, when he is excited. But I don't mind about that. I don't mind about any of that. Soon I will be alone in the 'home' here because the boy Billy Brent will leave for Backwater College in Virginia and the girl Adele will go to Sarah Lorentz near New York. They will be in the universities apart.

"Already in 1936 there is much snow. I go to sleep in blackness and wake up to a white world. Then I go to school in my galoshes. I sing in the school choir—tell everybody—but I think I will be forced to leave. I am on probation, and I cannot make a false move. It is very serious. (In America, there are strange books filled with filth that are forbidden in schools.) Also, the negro maid of Uncle Lester and Aunt Clara has paralysis agitans. It comes and it goes. Her name is Rhea. I am her pet when we have dinner there in the big house in Mt. Washington. (I can also spell *hysterectomy*, it is good to learn such words for English studies.) Sometimes in bed, late at night, I think about the negro maid, about *paralysis agitans,* but by the next day I have forgotten. It is that I am fickle. I do not stick to anyone. I am always looking for the next person. Even when Sally Krieger puts her tongue in my mouth, I am thinking about the next person. Who would want me, knowing that? Not even I want myself.

"Did I mention, by the way, that in Baltimore, a little boy, aged eight, was swept away in a sewer drain for two miles and came out alive at the other end! It is true. Six thousand seven hundred and ninety-four dollars were found in the clothes of a dead man lying in a downtown gutter! Again the truth. Four hundred Republicans from the state of Maryland came for dinner at the Lord Baltimore Hotel and shouted insults at President Roosevelt and the New Deal through each course. Also, tell the *tantes* that Mrs. Lyttleton Purnell gave a supper in the Valley in honor of her mother's ninetieth birthday, Mary Butler Shearer got married at the Catholic Cathedral, and the Christian Bachelors of Baltimore had a Cotillion at the Opera House where all the Christian ladies wore white dresses and made formal curtsies to the Christian Bachelors each time they danced together. All this the newspaper says.

"Of course, some forces are centripetal, too. Not everything flies apart. How could it be otherwise? It is not the apocalypse. Even so, you have to look where you're going, as in Frankfurt. That is only natural. It is all like a dream. I am in the same old trance. Sleepwalking. Yes, sleepwalking. Here strange things are happening . . ."

It was enough. Manfred sat back in his desk chair and rested. His writing arm ached, he had worked so fast. A vague hum of conversation came from downstairs, the fading domestic drone of the three Nibelungen, eating the last of their Jell-O in the dining room. Manfred heard Adele laugh at something her father said. When he tried to close his eyes against the sound, he could see a vision of Kurt shining in the middle distance, shaped there loosely, as though he was waiting for something to happen. He could see a vision of them all, waiting together in Frankfurt for something to happen. There was his father, waiting, his mother, smiling as though she knew a secret, and all the rest, including his aunts and uncles and his friend Joachim Adler. They were all shining with an old familiar light. But while his heart expanded like a football as he sat at his desk, the light began to waver. It narrowed slowly to a pinpoint, flared for a moment, then vanished. Now it was Manfred who waited. All he had was his long skinny room and the renewed darkness behind his eyelids. Blinking, he forced himself to read over what he had written. The act gave him solemn pleasure. His letter contained only a single joke, maybe two. It was how he felt. It was exact. Manfred Vogel loved exactness and clarity; sanity, as well. He closed his eyes again, but could see nothing. He was too tired, the whole world was too tired. He told himself that there wasn't much time, without quite hearing his own words, that time was running away with itself, on one of its familiar rampages; then, after waiting some more, he tore up his work into halves, into quarters and eighths and so on, hardly knowing what he was doing, finally getting up to flush the tiny white flakes down the toilet. That was the end of Manfred's evening. Downstairs, meanwhile, Adele was still laughing at another of her father's punning jokes.

Part Three

Chapter One

He took to sleeping over at the Ottingers' on Friday nights, after the Gordons had left for Fairfax Road following the ritual weekly dinner. Prayers, Sabbath lights, *hallah*, sweet wine: it was as piously cloying, as tiresome as Saturday morning in the synagogue (where Max had taken him once or twice to show him off to the congregation), and Manfred rustled impatiently through the ceremonies until the last singsong blessing was done with and they were all released for the evening. Later, Stewie's vast bedroom awaited him upstairs. His cousin's adolescent cave was stuffed with memorabilia and magical family debris that went back to the Civil War and even earlier, dim Ottinger accumulations passed along like icons from one forgotten generation to another. Nothing was ever thrown out, nothing could be moved. That was Family Law. There was a three-legged milking-stool, used a century ago when the family kept a cow in a backyard downtown; a Bible in four colors, copied from an elegant Soncino version; a framed wedding invitation, date blurred but established by family tradition as 1852, uniting Betsy Friedenwald and Elie Ottinger; old sepia postcards from Atlantic City stuck in the mirror over Stewie's dresser. There was also a tarnished brass bed in which Lester's twin uncles, Marcus and Simon, had been born; Lester's own baby shoes, saved in the Great Baltimore Fire of 1904, when the Ottinger house on Paca Street burned to the ground; and an early-model steamer trunk, bearing a sign that read, "Packed at Home, Carried Everywhere." There was nothing from the Gordon side, however, there having been nothing to bring to Mt. Washington when Clara Gordon married Lester, beyond her Goucher diploma.

From the windows of Stewie's room, Manfred had a sweeping view of Pimlico Racetrack across Rogers Avenue. Sometimes on Saturday morning, a few horses worked out just after dawn, racing lightly against themselves while the sun rose dimly over the Green Spring Valley. Nothing else moved then, except for the rippling flash of the horses. A blush of mauve light suffused the sleepy track, billows of white cloud moved slowly with the sun, overhead a huge arc of softly bleached new sky capped it all. The stately Ottinger firs stood guard in front of the house, a row of budless hedges bordered the pavement. Manfred had never felt such stillness. It was the stillness of the American winter, unlike Frankfurt, unlike anything he had ever known in Germany. It was a solemn, cushioning presence inside the big house, too, heavy, protective, and indisputably adult (perhaps Clara's and Lester's finest creation).

Things were different on Fairfax Road. There the abrasive struggle with life began as soon as everyone was awake. Sneezing, coughing, flushing sounds; Max's gargles, flatulence; cataleptic radiators resisting heat; Adele's quiet but assertive morning voice; Florence's boiling pots. Life was atomized at the instant of awakening, centrifugal forces took them all by the throat. Manfred told himself that it was hateful, compared to the way Clara and Lester chose to live. Here in Mt. Washington, across from the sweating horses, Manfred had a bathroom of his own, almost as large as his skinny bedroom on Fairfax Road. Boxed into the ceiling was a slanting skylight with a single pane of glass, like an eye to heaven. Fallen leaves stuck to the pane. Once, standing inside the bathroom with the door closed, Manfred tried shouting Adele's name at the top of his voice and discovered that no one could hear him. It was the stillness again, an Ottinger masterpiece of privacy and privileged containment.

After dinner, once the Gordons had left, the trio from *Rosenkavalier* twined its scratchy way up to the second floor from the sunroom, sounding slightly sharp on the Ottinger victrola. It was Lotte Lehmann and her pals again, singing in Manfred's own language for Manfred and Lester. The three women's voices ran together like shining metal, twining upward. Manfred stopped to listen, resting on Stewie's bed. "*Hab mir's gelobt . . . ,*" he heard, in the Marschallin's resigned voice. Tears of self-pity rose to his eyes at the sound of the words, surprising him. It was the nostalgia again, the sudden tug of his mother's smile rising with the music, the secretive smile sometimes reserved only for Manfred and Kurt. (Nostalgia was like eczema. It made you bleed.) "There are many things in this world we never believe exist, even when the evidence is clear . . ." Lotte Lehmann sang. (He didn't have trouble believing, he had trouble accepting.) Downstairs, Uncle Lester sat in his cracked leather chair, hand cupped dreamily to his ear, sipping eau-de-vie alone in the dark as he listened to the trio. It was his favorite. To Manfred, Lester looked worn out at the end of the week, his forehead grooved with commonplace worry lines, just like everyone else, his eyes paler and more faded than usual, even though he didn't have to go to work or do anything ordinary. He looked like Max Gordon or Julius Vogel. As Manfred and Lester listened, a floor apart, the record played on to the end. The mood grew lighter, a German waltz began. Manfred began to hum to himself. The little blackamoor ran in to retrieve his mistress's handkerchief. The woodwinds spiraled wittily upwards as the curtain came down. It is all a joke, they said, it doesn't pay to take life too seriously. But Manfred knew that already, had known it for a long time, and besides, it didn't seem to matter how you took life, life made all the rules, anyway. Then the opera was over, the house was quiet. Lester came upstairs to brush his teeth and take a sleeping pill, Manfred could hear him shuffling down the hall, and by ten-thirty they were all in bed: Rhea and Shorty safe in their apartment over the garage, Lester and Clara side by side behind closed doors, sunk deep into imported goose-down, and Manfred in Cousin Stewie's brass bed, trying to read *The Jew of Rome*, at Clara's urging. Alongside the dresser, a Forest Park pennant hung limply on the wall, faded green and grey. A half-empty jar of pipe tobacco sat on the dresser top, beneath a photograph of Stewie's Pi Tau pledge class. There was also a rack of old pipes. Within a half-hour, only the nightlights in the upstairs hallway were still burning, the etherlike silence had anesthetized the house, and Manfred had already fallen asleep over Herr Feuchtwanger's all-too-steady prose.

On Saturday morning, they sat together in the sunny breakfast room, as though they were a real family, sharing orange juice, scrambled eggs, and *hallah* toast prepared by Rhea. Shorty did the serving. Over coffee Uncle Lester would quiz Manfred in a sleepy voice, a continuing pastime that Manfred had grown to enjoy because he was good at it. Hum me the tune that Cio-Cio San sings when she thinks of her lover . . . Who is the governor of Maryland? . . . Describe the separation of state powers in the USA . . . List the tribes of Israel . . . What was Verdi's last opera? . . . Is Bruno Hauptmann guilty? . . . The sun streamed in the windows, the coffee boiled up in the morning light. In the kitchen, Rhea's hand trembled behind her back, where she often hid it now, and in the breakfast room Clara choked on her special medication for iron deficiency. Governor Nice, Manfred said to Lester, awaiting approval. Lester nodded. The tribes of Israel are Dan, Reuben, Judah, Manasseh . . . He couldn't remember the rest; it was tiresome trying to get the tribes of Israel in order for Lester. Falstaff, he answered. Hauptmann is guilty, without a doubt. Clara shuddered with the effort of swallowing another pill. At breakfast the *Morning Sun* papers lay folded alongside Lester's place setting. His napkin was stuffed chaotically into its ring. Rhea could no longer handle chores that demanded neatness and order.

"We're going to see some bloody action before this thing is over," Lester commented a few minutes later at the news he was reading. He left the remark unexplained and, coughing away morning phlegm, turned the page. His pajama collar was skewed over the top of his bathrobe. A wiry stubble covered his chin, there were sleep marks on his cheeks. Reading on, he began to speak to Manfred at the same time. "You giving any thought to college yet?" he asked, scanning the page casually. "Don't forget Penn when you do. We've got pull there. Stewie's there and I was there. My cousin Freddy Gunstermacher went there, too. Where do you think you want to go?" He put the paper down and looked into Manfred's eyes. "You have to go to college, you know. In this family, we all go to college. You understand that, I'm sure. Tell me, what's the nature of your problem at home? What are you trying to run away from?"

It was one of Lester's tricks. He talked on in an even tone of voice, hardly letting the pitch change from one sentence to the next, then a hand grenade blew up in your face, just like the one the young man had picked up so innocently in his backyard. Manfred reached for a piece of cold toast.

"Leave the boy alone," Clara said in French.

"Am I bothering you?" Lester asked, facing Manfred.

"*Ça va bien avec moi*," Manfred answered.

"You sure?"

"I'm all right."

"Actually," Clara said, "if you didn't have a problem, given circumstances, matters would be decidedly queer, don't you think?"

"Max tells me there's something on your mind these days," Lester persisted.

"Don't be shy," Clara said.

"Nothing's on my mind."

They stared at each other without expression.

"Well, almost nothing. Maybe I'm a little bored," Manfred said, turning away.

"You're too young to be bored," Clara answered. "I don't take that seriously at all. I never believe young people when they say they're bored. They always mean something else."

"You don't come to Hagenzack's with us anymore," Lester said.

"That's what I mean. It's boring at Hagenzack's. They don't let me do anything there."

"It would be perfectly natural to be homesick, you know. If that's what it is."

"To this day, even unto medical school," Clara said, "your cousin Stewie is homesick in Philadelphia."

"How's your mother doing?" Lester asked.

"My mother's fine."

"What does fine mean? Your mother just had a serious operation."

"They tell me she's fine. She wrote me herself. She's home. She's doing fine. All it is is she can't have babies." His mother's letters were formal and solicitous; they also scolded; they told him to behave himself, not to be a burden and embarrass them in Frankfurt; they said nothing about not having babies, about anything real. "Listen," Manfred said, "I promised to call Jojo McAllister." Another little lie.

"Is Adele getting on your nerves?"

"Oh, Adele."

"Is she?"

"Adele? I mean she's right there, all the time, in the room next to mine."

"So she *is* getting on your nerves."

"That one could use some controls," Clara said. "I always knew it."

"I promised Jojo McAllister to go to the movies with him this afternoon."

"Nobody's stopping you."

"Well, I promised to call him."

"Son, it's not even nine o'clock," Lester said sadly, checking his watch.

After a moment, Lester gave Manfred a last quizzical look and, stroking his chin doubtfully, returned to the paper. Clara then poured herself a second cup of coffee and took it upstairs to bed. It was silent again in the house.

It was also perfect, in the Ottinger way. The breakfast room was all yellow. The remains of Manfred's eggs were yellow, so was the butter and the wallpaper. It all glowed in the sun, in perfect yellow unity, Clara's and Lester's plan. "Well, I guess I better get a move on," Lester said, rubbing his beard again. "Unless you feel like a little postbreakfast chess." "Maybe later," Manfred said. "As you like," Lester said, heading up the back staircase. "And don't keep Jojo McAllerwhom waiting too long," he added in an acerbic voice. Then he disappeared overhead and Manfred was alone again, the way he wanted to be. He examined the yellow wallpaper, covered with spring daisies, the yellow napkin spread across his lap. They gave off a quiet scent of money and Ottinger permanence. Their luxury soothed him. Luxury, it turned out, could make up for a lot of things.

Upstairs, later, Manfred spent fifteen soapy minutes in the beautiful green shower stall in Stewie's bathroom, steaming the place up. Raising his head, he could hardly see out of the skylight. Then, after putting on his clothes, he left the house to take a walk, heading down to Green Spring Avenue, then cutting over to Park Heights, where the trolley tracks ran in the gutter, right next to the pavement. It was still country out there, almost, spotted with hilly farmland and great homes sloping down to Park Heights Avenue on icy lawns. It was certainly better than the narrow street and the faded brick shack on Fairfax Road, it beat the squeezed-in little rowhouse by a mile. It almost made it look tawdry. Also, out there, in the country, there was no Max Gordon, who had taken to calling Manfred "The Old Probationer," in a typical facetious effort to smooth things over, no Florence, walking around her tiny kitchen with one hand resting on her belly as though trying to keep it nice and warm, urging Manfred as she worked to be nice to her husband, to let bygones be bygones and pick up their forgotten rummy game again, no Louis Lambrino, slouched under the unsupportable weight of countless umlauts, no Athol Burns, fuck book in hand; and no Adele Gordon or Billy Brent, either. The last was probably best of all.

Chapter Two

After the holidays, the Saturday night parties continued in neighborhoods all over the northwest corner of town, sweet, dipping, musical adventures below ground in which Manfred had become expert. He now knew exactly what was expected of him when the jukebox came on. His forearm braced the girl at the waist, his left hand just as firmly clasped her right and set the direction, and as the piece slowly played itself out, he pressed on, bending his willing partner backwards. Thighs, crotch, and bellies met, paused a moment, met again. Manfred had it just right; so did everyone else. In Joanne Lerner's basement, in Toby Miller's and Gloria Levitsky's, in the basements of dollies everywhere, the world shook with cosmic molecular collisions that might go on echoing for generations. On the spot, genetic fate was being decided. One dip could change the course of a life and transform the future.

By now, Adele was confidently rotating among the three young men she had fallen in with over Christmas, a different one each Saturday night; but such inconstancy didn't last. Within the month, it became clear to Manfred that Saturday night was now being saved—and no questions asked—for Bernard P. Potter, who had clearly been her favorite all along. Eliminated were Len Schwartzman, who had tiny yellow teeth and liked to shoot craps at Saturday night parties, and Sammy Billig, who between dances, at which he did not excel, wanted only to talk about Karl Marx and the Workers Manifesto. All of this Adele dropped in breathless bits and pieces at the dinner table, with a new self-satisfied edge that made Manfred squirm in his chair. Len Schwartzman was actually sort of sweet, Adele told them condescendingly, he was like a little clumsy puppy when you got him alone, even if his discolored smile almost froze you in your tracks, and Sammy Billig, while full of brainy energy and an impressive vocabulary, was really quite boring. As soon as he walked into the house, he insisted on engaging Max in desperately earnest conversation about politics and hardly seemed aware, at least as

Adele saw it, that she was planted on the blue rug right next to her father's chair, all dolled up, ready for a night out. Neither would do, she made it clear, Len or Sammy.

Bernard P. Potter was another matter. B.P. was as tall as Billy Brent. His family had just moved from Cincinnati to Baltimore. B.P. had perfect teeth and a limited number of words at his command. This last probably came from a foreshortened imagination, Manfred thought, unkindly. Like Billy Brent, B.P.'s head was covered with tight curls, but black and shiny in his case. Each one looked as though it had been wrapped around somebody's little finger and set permanently. Was he handsome? No, Manfred decided over Christmas, he was merely tall. His nose was long and bony, he had large ears and full lips, and like cousin Stewie, and Florence and Max, he was fond of tobacco. Manfred could smell the cigarette smoke on his breath as soon as he walked into the house, as powerful as the ashy bottom of a coal furnace. The smell was in his clothes, too. B.P. lived in Windsor Hills, a mile away, and he was killing time, after the move from Ohio, before going off to the University of Virginia next semester.

"I think he's darling," Adele said, in a strangely affecting voice, when asked. Manfred had never heard her sound so wistful and so uncritical, too. And Florence agreed about B.P., in her own way. "The thing is he's got style," Florence said, finishing off her prescribed glass of milk after dinner. "Good manners like that don't grow on trees. So he's not a genius, who needs geniuses, just look at the Billig boy, all that talk about Russia."

Max didn't object to B.P., either. B.P. reminded Max of good things, success, for one. B.P.'s father had been brought east to manage the May Co. The May Co. was part of a national chain. Success there was success anywhere. Potter père had to be good at his work to get a position like that. And Max was beginning to enjoy success, too, the first flourishes already surrounded him. Already, as the profits slowly grew at M. Gordon & Son, Max had hired extra sales help; already, he was eyeing the space next door on Lexington Street for expansion; already, he was seen strutting naked around Hagenzack's with his chest stuck out; already, he was planning where the new Gordon house would be custom-built, once the baby was born. In another five years, well, who knew? America was America, as it was always reminding him, anything could happen.

Yes, Max liked to have B.P. around. He said as much to Adele: gentlemen like Bernard Potter were always welcome in the Gordon house—this spoken with still another significant glance at Manfred. (Since probation, life on Fairfax Road was filled with significant glances.) Such condescension naturally irked Manfred. It made him even more jealous of B.P., made his behavior around the house seem downright peculiar at times and occasionally inexplicable. He was beginning to take himself by surprise, he discovered.

One Saturday evening, for example, around eight-thirty, just when B.P. was expected to show up in his Chevy convertible, Manfred was standing at a window in the passion den upstairs, watching Billy Brent pace the pavement outside, pretending to be taking the air. Like Adele, Manfred was getting dressed to go out. At the moment, he was buttoning his shirt and taking his time about it. Sally Krieger awaited him in the basement of Peggy Lyman's Tudor mansion out on Seven-Mile Lane. So did Alice Sacks—he hoped—who just last week, dancing with him, had thrust her thighs into his groin without warning.

"I don't think we'll be able to make it out there tonight," Adele said, in an offhand way, edging alongside Manfred at the bedroom window while she brushed out her hair again. Electricity crackled with each stroke; in the dark room, her face was dimly outlined by the streetlight outside; her broad straight Gordon nose, in profile, tipped up surprisingly. Glancing at Adele shyly, not wanting to be caught, Manfred saw that her eyes were turned in on themselves, catlike, wholly absorbed in what she was doing, indifferent to him. Manfred tucked his shirt into his new long pants and looked nervous.

Again, he couldn't tell what kind of mood she was in. Good or bad, there was hardly a sign. It was always like that now. "You don't think you'll be able to make it out where?" he asked icily. As Manfred spoke, B.P.'s car turned the corner. On the sidewalk below, Billy Brent stopped his pacing.

"To Peggy's." Adele peered out the window, hands at her side. Manfred saw a shiver of silver happiness wash over her face as B.P.'s car came up the street. B.P.'s car had a soft cream paint job, black top, and wheels with white spokes. There was nothing like it on Fairfax Road. At the approach of B.P.'s car, Manfred could feel Adele's excitement grow.

"And where may you make it to?" he asked in a mannered voice.

"Someplace else. Another party. You don't know them."

"Well, I guess I can take the streetcar out to Peggy's. I'm a big boy." Manfred waited a moment before going on. "Tell me," he said, "what kind of name is Potter anyway? I don't think he's even Jewish."

"Of course he's Jewish. What makes you think he's not Jewish? What are you getting at anyway?" She glanced sideways at him, brush threateningly in hand. "Come on, tell me."

"I just thought," he answered. "I never heard a name like Potter."

Adele shaded her eyes against the glare of the streetlight and stared out the window. "Well, now you have," she said.

B.P. was backing his car into a space across the street, in front of the McAllister house. The white spokes glittered as they turned in slow reverse. B.P.'s car made the only sound on the street. "God, it's gorgeous," Adele breathed in a distracted whisper.

"That old thing?"

"You only wish you had one like it."

Manfred smiled. His whole body suddenly expanded. "Do you love him?" he asked, pointing down at the street, where Billy Brent was leaning against a sycamore tree. "Love who?" "Him," Manfred answered, deliberately vague.

"Do me a favor, please, and mind your own business."

"Come on, Adele."

"It's one of your worst faults. Sticking your nose in my business all the time."

"I care about you, that's why. You can tell me, Adele. Do you love him?"

"Manfred, cut it out."

"Do you love *me*?" he then asked. "Come on, say it, Adele. Say it." He hardly recognized the sound of his own voice, or the insistent whine that came through each word. Somebody else seemed to be doing the talking.

"Are you trying to be cute or are you just acting nutty? Because if it's nutty . . ."

"I want to know."

"You're crowding me, Manfred, and I don't like it."

"Adele?" It was Florence, shrieking from downstairs. "He's here on the porch."

"He? I don't know any He."

"Lo-chin-var." Pronounced phonetically.

"If you want to be a friend," Adele said to Manfred, "get downstairs and keep her out of sight. Distract her. She'll spoil everything."

"And what are you going to do about *him*?" Again he pointed at the shadowy figure guarding the pavement below.

"Him," she said, stretching the word out. She lifted a hand as if to blot out the view. "Ruler of All He Surveys. King of Beasts. He better not make trouble."

"I thought you said you loved him."

"I never said anything about love. I never said anything about him. He's just a nice neighborhood guy. Okay, a little prettier than most. I like him, everybody likes him. But ordinary, plain ordinary. He'll just have to learn. I mean you don't have to be Einstein to learn. He's no better than anybody else. I'm ready to abdicate anyway. I already have, but he won't believe me. Laura Piscitelli can have him. They make a good pair. Robin Hood and Maid Marian. Tarzan and Jane. She's dying for him." Adele paused a moment and turned on Manfred. "And so are you," she said, mercilessly. "Everybody knows that."

He didn't wait after that. He grabbed her, his two hands framing her face. "No," she said, trying to pull away. But for once, he was too fast. Holding on, he leaned forward blindly and began to kiss her on the mouth, in a sudden panic, tracing the curve of her full lips with his tongue, round and round, purring with pleasure at the same time. It was how he always kissed Sally Krieger and the other dollies, how they had taught him. Holding her head like that, he could feel the unexpected strength of her cheekbones under his thumbs, their beautiful symmetrical curve, the softness of the flesh over her temples. His mouth opened; he knew what he was doing. So did she, even though she held back, resisted, even clawed halfheartedly at his upper arm with her hairbrush. Adele, he was thinking all the time, as though he could contain the whole world in the sound of her name. He was suddenly seized by hope and joy, twin strangers. Adele . . . There was a muffled sigh, a peculiar breathy sound from one of them. After a while, when Manfred let go, both of their bodies went slack.

"There," Manfred finally said, as though he was giving her back to herself. "I had to do that. You know?"

Adele had dropped her hairbrush. She bent over to pick it up. There was suddenly a lot of conversation going on downstairs. "Look what the wind blew in," they heard Florence shout. Straightening up, Adele groaned. "Hello there, young man," Max said from his easy chair. A match struck, a cigarette lighted up. Upstairs in the passion den, Adele and Manfred moved apart. Out on the street, in the dark, Billy Brent disappeared.

"I had to," Manfred said again, his eyes fixed on Adele.

"You had to," Adele mimicked, turning away. "What do you know? You're not even out of the cradle yet."

"Don't talk like that."

"I thought you were different, not like the rest of the jerks around here. Special, that's what I thought. Some kind of gentleman. I thought you had some manners."

"I'm going to kiss you again."

"No, you're not. Just keep your distance. Stay away. Listen, this never happened. Remember that. It never happened. I swear, I must be going crazy or something. It's your fault. Everything's your fault."

"Adele," he began.

"Adele nothing."

"Listen to me. Please."

"There's nothing to say. I'm late now. And he's waiting downstairs. I have to finish getting dressed. The whole thing's crazy."

"Just listen to me," Manfred said. But he had no idea what he wanted to say.

"Go on downstairs now." She pushed him towards the door. "Distract her. Help me. Please. And I'm sorry for what I said before. About you and Billy. You know me, I forget myself, I lose control, you can't believe a thing I say."

Later, after Adele and B.P. Potter had left, Manfred took out his navy beret and wore it to Peggy Lyman's party. He wore it on the number 31 trolley and then on the number 5, when he transferred out in Arlington. The boys at the party all hooted when they saw it, but some of the girls pretended to be enchanted. It seemed so European, so true to Manfred's cosmopolitan nature, so original. (What did they know out on Seven-Mile Lane?) At one point in the evening, Alice Sacks pulled it over her head and pretended to be Manfred's Apache partner. She slithered along in Manfred's arms, exaggerating each gesture. She thought she was doing the tango. From the sidelines, Sally Krieger made nasty remarks and looked disgusted at the whole ridiculous charade. Manfred himself was on the edge of hysteria. He could feel it when he laughed. Something ran in his bloodstream, like a shooting star. "A couple of dopey freaks," was Sally's judgment on Alice and Manfred, repeated aloud a half-dozen times so everyone could hear. When she finally said it to his face, just to wound him, Manfred had to agree with her.

In the course of the evening, Manfred danced seventeen times in a row, took a rest, then went on with the marathon. He never really got tired, he seemed to have endless strength. He danced with Alice, Sally, Nina, and Louise Binswanger, who put her pride behind her and twice asked him out on the floor. At midnight, there were tiny hot dogs and potato salad to eat. By then, the trolleys had stopped running, so Louise Binswanger and her date gave him a lift to the corner of Liberty Heights and Garrison. From there, he made it home on foot, beating Adele and B. P. Potter—who never did show up at Peggy Lyman's—by at least a half-hour.

Chapter Three

Manfred had discovered a new word, or rather he had suddenly discovered the meaning of an old one in terms of himself. The word was exile, both noun and verb, subject and act, and it had leaped off the page in an editorial he was reading in the *Evening Sun* about the question of German refugees in the United States. Egg-zile, Manfred read, extracting all the weight from the two syllables. There seemed to be serious doubts on the matter in America. The printed discussion, with its self-congratulatory headline, "The New Eden," shocked him. How many refugees to admit? How many exiles to sustain? Who would pay?

I-am-an-exile, Manfred said to himself, translating Carl Schurz's memoirs aloud in Mr. Lambrino's class. So was Carl Schurz, he reminded himself. I-am-an-exile, Manfred breathed over and over, romantically mouthing the mournful new word at his Indian desk at night (listening for sounds of Adele), repeating it inside his head as he sat in study hall pretending to work, or while he checked in with Athol Burns every week, to reconsider again the touchy question of probation. Albert Einstein was also an exile. And Heinrich Mann, from Lubeck, Ernst Toller, and Manfred's Friday night companion, Lion Feuchtwanger. They had all had to scramble shamefully to stay alive. They had all had to run away from the Homeland, like Carl Schurz. The sanctity of the State demanded it, the purity of the People was at stake.

"Herr Vogel, you're drifting again." It was Louis Lambrino's languid voice, exposing the obvious. Manfred's own voice wound down. He looked up sleepily, marked the place in Carl Schurz's remembrances. Carl Schurz was a great German liberal from another time. There was something in the tone of his voice that confirmed it. He was always optimistic. Life righted itself at the end of each chapter of his memoirs. Life was progress eternal. But at the moment, in class, Louis Lambrino was at Manfred again. He was at Jojo McAllister too. Louis Lambrino couldn't control his loathing for Jojo

McAllister. It caused him to pinch the bridge of his nose between thumb and forefinger and turn away from the class in pain. It made him roll his eyeballs into his skull. Once, in a rage, he pretended to tear his hair out. Such displays of furious contempt in Authority filled Jojo with guilt and terror, they filled all the students with guilt and terror; the class trembled at their desks in front of Louis Lambrino.

Every Wednesday morning Manfred got up an hour early to coach Jojo in study hall before school began. "I can't do it," Jojo protested from across the aisle. "I told you a hundred times." Manfred reached out and stroked Jojo's forearm solicitously. "Everybody else is able to do it," Manfred said in his most soothing voice. It was a relief to be in a position to give someone else a little advice for a change. "I did it. You can, too." "What are you talking about, you did it, you're a Heinie already. You were born German. And quit feeling me up." "Now, slowly," Manfred said. "*Der, die, das, die.* It's easy."

During class itself, while Authority stood at alert in front of the blackboard, Manfred was always whispering answers to Jojo under his breath, short, sharp, rebellious ejaculations of one or two syllables. He slipped him scraps of irregular verb conjugations during tests, shivering at his own boldness, once he even scrupulously inked a vocabulary pony on the inside of Jojo's wrist that got Jojo an amazing 72 on a quiz, raising Mr. Lambrino's expressive eyebrows a full inch in disbelief as he toted Jojo's mark again. But victory was brief. There could be no happy ending to this chapter. Jojo was failing, everybody knew it. It was horrible, but it wasn't horrible enough for Louis Lambrino. Unconditional surrender in public was what he demanded, almost impersonally. He thrust his pelvis forward. His lower lip pouted. The sardonic comments flowed easily in the face of intellectual shabbiness. "Would you be so kind, Herr Vogel," Bigprick might begin, before slicing one of Manfred's veins so cleanly that Manfred hardly noticed the pain at first. "And as for you, Herr McAllister . . ." Mr. Lambrino pinched his nose. Another nail found its way into Jojo's coffin. A master: it was clear from the beginning. No one else at Forest Park could quite manage such stylish malice (and authentic despair) in the classroom.

By February Athol Burns had placed Manfred in a regular class, although probation still held. Manfred was now an almost-conventional member of the student body at Forest Park High School, one of one thousand fifteen hundred and seventy-three. "You understand your responsibilities, I assume?" Athol Burns asked from across his desk.

The exile nodded.

"You do or you don't?"

"I do."

"You have a habit of speaking in sign language. It's unnatural. You signal instead of verbalizing. Are you afraid of words?"

Manfred didn't know. Words, spoken aloud, might kill. He had had that experience before; or close enough. You always had to watch what you said. His parents had been telling him that for years and were still at it in their letters; everywhere you went, his mother wrote, the walls had ears. He clutched the books in his lap.

"You see? Your mouth is turning down at each corner, right and left, thereby offering me an unmistakable clue to your true feelings. I know all the signs." Athol Burns was a fervent practitioner of physiognomical diagnosis, of the easy-to-read variety.

Manfred tried to gather his true feelings. He had trouble recognizing what they were sometimes. In fact there were no unmistakable clues to them. What did he feel for Adele Gordon, for example, passion or friendship? How had he felt when he grasped her face between his hands and kissed her? He didn't seem to know from one day to the next. (But he called it love and was momentarily satisfied; the stirring memory of hope and joy raised that evening in the passion den persisted.) When he was being challenged, as he was now, in the assistant principal's office, feelings whirled through him like splintered glass. In the morning, just between orange juice and coffee, he could live through endless manic transformations as he faced Adele and her parents across the breakfast table. Crazed heartbeats echoed at a mention of the new baby. Unrecognizable emotions resonated in the small room when Adele described her plans for a weekend with B.P. or casually mentioned Billy Brent. Headaches came and went with each remark, then returned at night, fearfully intensified as he read and reread the noncommittal letters from Frankfurt or looked again at Kurt's picture, plucked from its place of honor in his wallet. His mother's violet handwriting caused pain, Kurt's face was sometimes the face of a stranger now. Hour by hour—like this very moment seated across from Athol Burns—a kind of demoralized inner chaos spun itself out at the speed of light, whatever Manfred did.

"I would be dealing less than fairly with you," Athol Burns went on, "if I didn't acknowledge your cooperative attitude in the recent past. It's gone a certain way—not a long way, not even a considerable way, but a certain way—towards revising my ideas about you. You yourself can take credit for that."

"Thank you, sir. I am happy to hear that, sir."

"Frankly, following the unpleasantness of our initial encounter, I have come to believe that you're really quite an intelligent young man. Not exceptionally intelligent, not quite that, but intelligent enough. And intelligent enough can pass for quite a lot in this world. You suffer, however, from exaggerated notions of your own importance. It is not a rare disease, as we all know, especially among the young, and I trust that you will learn soon enough that life-experience is the only vaccine that can immunize the spirit against it. I detect, we all detect, even the teachers who admire you the most, a certain conceit in face of the world, a callowness in your makeup slightly beyond the norm. Slightly, in this case, is as good as a mile. Our students would call it stuck-up. Do you understand stuck-up?"

"I've never heard of stuck-up."

"You've never heard of stuck-up?"

"So far the word has escaped me."

"The definition is somehow encapsulated in the tone of your last statement."

"Sir?"

Athol Burns gazed at his ward. His copper skin turned a little darker. He ran his tongue over his teeth. "Sometimes, my friend, I don't know how to take you," he said. He paused then to chew on a fingernail. "Be so kind as not to bait me. It'll take you a long way. That's fair warning. And I don't give many warnings. Well, so be it, I hear that you're tutoring for Miss Klinesmith."

"She lets me take over the class sometimes."

"It's an interesting experiment. We'll see how it goes."

"I'd like to take over German studies, too. From time to time, that is."

"Not a chance. Mr. Lambrino owns that territory. What's his is his."

"I think I know that, sir."

"I doubt that he'd ever let any authority out of his hands."

"I'm sure you're right."

"You seem suddenly very agreeable."

Manfred made a self-deprecating face.

Mr. Burns shifted in his chair, then folded his hands on his desk, like Miss Ficker in the office next door. "Miss Feinstein," he went on, "also says you're doing outstanding work for her."

Manfred bowed his head.

"Why so abject?" Athol Burns asked, leaning forward impatiently. "With you, it's either-or. I'm not knighting you, for Chrissake, I'm passing along a favorable opinion."

"Thank you, sir." Manfred faced into the weak sun that came through the window behind Mr. Burns, squinting each time he had to respond. Wispy clouds scudded overhead in front of the sun.

"You seem to be turning into something of a specialist in contemporary history."

"It interests me."

"Miss Feinstein gave me your paper on the Versailles Treaty. Good work, far better than we're used to here. And it is salutary to read about such questions from the enemy's point of view."

"Ah, yes," Manfred said after a moment's pause for thought.

"You do bring something special to our little suburban high school."

"I'm grateful that you think so."

"You're not run-of-the-mill at all."

"Yes, I know," Manfred said. "I'm weird."

"I didn't say anything about being weird."

"Isn't the enemy always weird, sir?"

"That's from your mouth, my friend, not mine."

"With due respect, sir, I thought I heard you say enemy."

"You're not trying to bait me again, are you?"

"Bait you, sir?"

"Because the shit will hit the fan if you are. I don't take to that kind of stuff, you should know that by now. Don't ever try to put anything over on me. I'm known around here for my bite." Mr. Burns made a clicking sound with his teeth. "Now look me in the eye, quit staring at the floor like that."

"Yes, sir."

Mr. Burns waited a moment before going on. "You know," he said, "that you lack openness, for one thing."

"Openness, sir?"

"Our students have a kind of candor and openness that gives the school its own special flavor. It's our distinction. It's the first thing you notice about Forest Park."

"You're right, sir. I saw it at once."

"No one's ever told you that you lack candor?"

"Not as I remember."

"Well, you do. You lack candor and openness."

"I'm a little shy, they say."

"That's certainly a curable handicap."

"Carl Schurz was shy, too."

"Carl who?"

"Carl Schurz, he ran away to America."

"I'm not talking about Carl Schwartz, I'm talking about you."

"I just thought maybe."

"There is certainly a sarcastic edge to your voice, Mr. Vogel, and it doesn't suggest shyness."

This time Manfred was silent.

"Don't try to take advantage of me."

"No, sir."

"You sure you heard me?"

"Yes, sir."

"Because, as I say, the shit will hit the fan, sure as we're sitting in this office on a Tuesday morning. I like the students at Forest Park to hear what I say."

"They do hear you, sir. They do. I hear you. I hear you all the time."

"I like students to pay attention."

"Oh, they do, sir. They all do."

"I'm not talking abstractly, I'm talking about you."

"Yes, sir."

"You know something, I don't think you're so shy. If you were so shy, you wouldn't claim you were shy. That goes with being shy. Shy people can't talk about themselves."

"Yes, sir."

"What I think is you're probably subversive. That's probably the word for it. It goes with a certain kind of intelligence. Negative. Pessimistic. Anti. Yes, you're subversive, most likely without conscious intent."

"Without conscious intent, sir."

"Can you hear the tone of your own voice? Listen to it. It's unwholesome."

"Unwholesome, sir."

"Oh, Lord," Athol Burns said, looking up at the ceiling. "Help me to help this one flee from unwholesome things."

"Yes, sir," Manfred answered.

"It will take work." His eyes were on Manfred again.

"I know how to work."

"It will take persistence."

"Yes, sir."

"Are you persistent?"

"I am persistent."

"You've got the will for it?"

"I've got the will."

"We'll do it together. Two together quadruples the force. A pair is worth more than twice as much as one."

"I know that, sir."

"We must have a program."

"I like programs, sir."

"You'll have to follow me."

"I'll follow you, sir."

"Brass tacks now. For starters, I want four papers out of you, each three hundred words long. Due Monday morning, first thing, in this office, for the next four weeks. Be prepared to argue your convictions. Be prepared to support your reasoning, stand up for what you believe in. Then we'll check progress, week by week, see where we are, where we're going, if anywhere. The first theme is "The Effect of Character on Action." Write it down. You'll never remember it without writing it down. Writing maketh the exact man."

Manfred opened his notebook and made a note: The Effect of Action on Character. 300 wds. Monday. Burns. He closed the book, stuck the leaky pen in his shirt pocket. Where had Athol Burns hidden Billy's delicious fuck book? Then Manfred began to whistle.

"Stop that noise. It's rude. If you have nothing to say, keep your mouth shut. Your slouching gives you away, too. You're sitting on a single vertebra. You could snap in two, just like that. Up straight now. Like an adult. I don't allow malingering in this office. You've picked up a little credit recently, hold onto it. You're not out of the woods, you know. I've still got my eye on you. You could be brilliant as hell, I'm not saying you aren't, you could get the finest marks in the history of Forest Park, but character is what I count first and what the world depends on."

"It comes first with me, too, Mr. Burns, believe me. It's how I was taught in Frankfurt. I learned it in Frankfurt from Goethe."

Mr. Burns's eyes narrowed. "Gerter," he said after a moment. His chopper-face came closer. "To each his own. Here it's Abraham Lincoln. See to it that you take all that to heart."

"Yes, I will."

"Yes, what?"

"Sir. Yes, sir."

"We'll continue our little visits next week. Be prepared to struggle. Be prepared to defend yourself."

"Yes, sir."

"Anything to add?"

Manfred looked as though he was deep in thought. "When will my probation be over?" he finally asked.

"That's up to you. Of course."

"Up to me."

"It's all in how you perform. The choice is all yours."

"Probation is hard."

"Precisely the point. You modify your behavior, you'll see us change, too. You choose to become part of us, sincerely part of us, you'll see, we'll make room, we'll open up, we'll let loose like a fart in a pig's ass."

Outside, a horn sounded. The sun disappeared, it began to rain. Athol Burns blew his nose.

"Can I go now?" Manfred asked.

"May I, Mr. Vogel, may I."

It was almost time for Manfred's lunch-period. Florence's lettuce-and-tomato sandwich was dripping mayonnaise and tomato seeds in Manfred's locker. Athol Burns and Manfred nodded at each other. A look of monstrous possessiveness passed across the desk, assistant principal to student. By now, Manfred was thinking entirely in German. He spent half of each day thinking in German; it made him feel close to the Mendelssohnstrasse; he spent half of his time holding imaginary conversations with his brother, and his mother and father, just to be able to hear their voices. A fart in a pig's ass. He would have to tell Kurt. He would have to make it funny. He grasped his books and made for the door. It would take a full five minutes to clear his head.

It was as Billy Brent had said. When Miss Tubman wanted you to do something, you did it, the fewer questions the better. "I want you to sing over the melody, no words, a kind of loud hum in a false voice, ostinato," she said to Manfred.

"I don't have a false voice. Charlie Crimmins . . ."

She didn't hear him. She never heard objections. He found himself talking on, over the purple growl of her own voice, finally smiling at the hopelessness of it. "Listen," she ordered, sitting down at the piano.

"Listen," Billy Brent ordered at Manfred's side. He nudged him in the ribs, edged against him. While Miss Tubman played the melody, to set the mood and pitch, the feel of Billy there extended from shoulder to hip. Billy of course was taller, but his presence was a help. "They asked me how I knew . . ." Already, Manfred was producing the sound Miss Tubman wanted. Everybody in Boys' Chorus was nodding approval except fatass Crimmins. The boys stood in two rows on one side of the music room. When they sang, they held their hands clasped behind their backs like soldiers resting at ease. At the piano, Miss Tubman's pearls rippled over her bust. Her purple hair needed dye. Patches of dust grey had begun to show through at the part. Fatass Crimmins made a face at the sound of Manfred's voice. The thorny shape of envy prickled his mouth. He stood slightly apart, his skin as pure and lustreless as wax. "Pay attention, shithead," Billy Brent ordered Manfred. Miss Tubman sang on with Manfred. "Now really do it," she said, raising her hands high above the keys, as though they were dripping wet. "Do it a cappella. That's the test."

Fatass Crimmins squealed a little. He was an expert at a cappella. He was the one who always sang above the melody, falsetto.

"Well?"

"I can't do it without the piano," Manfred said.

"Jesus," Billy mumbled. "Just do what she says."

"I can do it," Fatass said, raising his hand.

"We all know you can do it, Mr. Crimmins." Miss Tubman swiveled on the piano stool. "Now, Mr. Vogel, ostinato."

Manfred inhaled. He tried to place his voice in the upper part of his throat, or, better yet, on the roof of his mouth. He wanted to make an unearthly sound. A quaver came out. He started over. Three flat tones tried to rise overhead. Billy Brent moved off to the side. There was a shuffling of feet in the Boys' Chorus. Aah, Manfred sang. "And what is that supposed to be?" Miss Tubman asked. Manfred said nothing. "Now, again." And she did him the kindness of striking a chord in the right key.

The struggle between Jerome Kern and Manfred Vogel continued for another five minutes. Those who hated Fatass Crimmins suffered intensely during the competition, he was sure to win by default. Manfred tried a second time. Then a third. He couldn't get the syllabic sound right, he didn't know anything about crooning, how could he? Somehow the syllabic sound was crucial to Miss Tubman. It was the smoke that gets in your eyes, she explained poetically. It was the ghost of unrequited love invisibly floating in space. It was the phantom of passion, uncurling itself painfully. It was what counted.

"Ok, Mr. Crimmins." Miss Tubman sounded weary. She turned a page of sheet music on the rack in front of her. "Your turn."

Manfred stared at the floor, unable to bear the sight of another human face. In front of him, a blackboard on wheels tilted forward at a slight angle. Notes covered a staff accusingly. Fatass Crimmins's voice coiled in the air. You couldn't tell whether it was a boy or a girl singing. Dead desire filled the room, ghosts and phantoms of love. How could Fatass Crimmins know about desire? Manfred peered around him. Billy Brent now stood a foot away, on his right. He was frowning to himself. The basses stood behind them. The basses all had to shave every day. They looked like Hugo Bondi. The room was beginning to smell like the cabin they had all shared on the *Europa*, a million years ago. Acrid odors floated alongside the sound of Fatass Crimmins's slick, androgynous voice. Manfred was growing sick to his stomach. Miss Tubman stopped playing. She held her head a little to one side, eyes closed in bliss or exhaustion, it was hard to tell. Fatass Crimmins went for the last note, a cappella. It was already five. Rehearsal had gone on for two hours. It was almost dark outside. "Aah," Fatass sang out in his soprano voice. Then, blessedly, Fatass Crimmins's voice broke at high tide. What a sound he made. What an ugly noise. Even Miss Tubman laughed. Pleasure flooded Manfred, pure satisfaction. Fatass looked as though he was going to cry. Whatever else, in the end the day certainly had had its compensations.

Chapter Four

Laura Piscitelli had taken to stalking him. He saw her out of the corner of his eye, a mere shadow, sloping along somewhere to the right or left in the school hallways, felt her purposeful leaning presence behind him in pursuit, at other times watched her drag her feet up ahead so he would be sure to catch up with her. She had that kind of walk anyway, slanted a little forward on her toes, as though she were on the lookout, body always tight in the torso, head and jaw thrust ahead of the rest of her, scenting the catch. At odd moments, too, she began to appear at his side just when he was about to open his locker. It was always a shock to see a girl in the boys' locker area. The two were kept strictly apart. Who could doubt then what Laura Piscitelli was looking for? At the sight of Laura, silently waiting there out in the hallway as though she were delivering a court summons, Manfred instantly forgot his combination and had to start all over again, mumbling the numbers under his breath for reassurance. He pretended not to see her, embarrassed in front of the other boys.

"Meet me on the track at lunch," she said one day. "We'll walk together."

"I'm busy," he answered, fumbling with the dial.

"Busy with what?"

"I promised Jojo to help him with his German."

"Jojo can wait."

"Sorry, Laura." He worked the combination carefully, trying to keep his hands steady.

"I'll see you at lunch, you hear? I have a couple of Hershey bars. We have to eat them before they go moldy." Then she skipped away without waiting for his refusal, her apple breasts bobbing in unison under her angora sweater. That day, at least, he managed to avoid her by hiding near the boys' gym during lunch break.

Walking home on Liberty Heights Avenue a few days later, she tripped him from behind. "Wait up," she called, sticking out her foot. Manfred lurched as she caught him on the lower shin. He was on his way to the newspaper vending machine on the corner, where the afternoon's headlines were visible by the time school let out. There he got a teasing report of the day's crucial events. The Rhineland question, as perceived by Germany, France, and England; rearmament; Japan and the Asian mainland; FDR. Especially FDR. FDR was everywhere in the headlines, every day, all week long. As for the Rhineland, it was clearly waiting for something to happen, something awful. That was what Manfred was looking for each afternoon, news of another soaring FDR triumph, news of war or European peace, news of Germany.

"I'm in a hurry," he said, quickening his step. His shin hurt.

"You're always in a hurry," Laura said, moving alongside him. She carried her schoolbooks like Adele, clasped in front of her breasts like armorplate. It made her round-shouldered.

"Go walk with Kitty Dean," Manfred said.

"I don't walk with Kitty Dean anymore. She never opens her trap."

"That never kept you from walking with her before."

"Oh, who's talking about Kitty Dean, anyway? I want to talk about you."

Manfred's eyes opened wide. At the syrupy tone of her voice, his vanity began to flutter at half-mast. So did suspicion.

"Do you like me?" Laura asked.

"Do I like you?"

"Oh, God, you're so slow," Laura said. She ruffled her curly poodle hair with one hand, in exaggerated frustration, almost losing her books in the process. "You're like molasses. They told me you would be like this."

"Like what?"

"That you play dumb. Adele told me. I see what she means. Adele says for a smart kid you can really be stupid."

"I have to go now."

"No."

They had reached the corner of Oakfield and Liberty Heights. Today's headlines could be seen through the glass of the vending machine. "Bipartisan Bitterness in Congress," Manfred read. That was disappointing. Manfred liked headlines with action, even if it meant disaster. At least, disaster meant that things were changing.

"Don't you get the paper at home?" Laura asked.

"Listen here, Laura, I'm busy. I told you a hundred times."

"My God," she said.

"You keep saying the same thing. My God, my God."

"Don't get all hot now," she said. She was wearing a plaid cashmere muffler wrapped three times around her neck. She hoisted her books close to her chest. They were covered with green oilcloth. "I think you're the best-looking boy I ever saw," she said.

The flag of vanity flapped stiffly in the wind, suspicion vanished. Laura stood there, smiling calmly. She seemed to have all the time in the world. Flushed with pleasure, Manfred began to read the weather report in the upper-right-hand corner of the paper. It was going to rain again, the report said, it was always going to rain in Baltimore.

"Well?" she said.

"You going home now?"

"Yes," she said.

"Come on, then, I'll walk you." And they proceeded to Fairfax Road in almost total silence, suddenly mute in each other's presence, parting in front of the Piscitelli's house without saying a word.

She finally nailed him on the track at lunch. Her cashmere muffler was wound around her neck. She wore her ugly earmuffs and a navy-blue pea jacket. The wind whipped around their shoulders, coming from the north. Manfred was shivering. "Here," Laura said, thrusting a chocolate bar into his hand. "Have a taste."

This time there was no escape. She had him, for fifteen minutes. Maybe she would flatter him again. He needed it. They wandered off together at an aimless pace. As they walked, Laura kicked little holes in the cinder track with the point of her shoe. Together they ate the candy, which came apart in their mouths like dust. "Here, you've got a spot of chocolate on your chin," Laura said. She reached up and wiped the corner of his mouth with her handkerchief. "Hmmm," she said, concentrating diligently. "Don't pull away so."

"I should be studying," Manfred said, enjoying her touch. "I have a history test this afternoon."

"Oh, we know all about you. You wrote the course."

"No. It's Coolidge and Harding and that stuff. I don't know that stuff at all."

"Well, don't look at me, I'm only up to the World War." She began to beat her arms against her body to keep warm. With one arm, as she beat on, she managed to brush hard against Manfred's side. "All that stuff is dead anyway."

"Only the actual events," Manfred said.

"What?"

"I mean only what happened is dead. The events are dead. But there's more than that to it. There are echoes, there are always echoes. Most people never hear them. But they're there, life is more than just what happened."

"Too serious," she said.

"Well, it's true. Nothing's really dead. Nothing's ever over. Especially history."

"Let's go have a smoke," she said after a moment. Manfred looked around nervously. There was hardly anybody in sight. "Come on," she said. "Behind the shop building."

"I'm on probation. You know that."

"No one'll see."

"Rigby's always nosing around outside his electric shop."

"I have some Old Golds."

"I don't even smoke."

"You'll watch me."

"I'd rather walk. I'm cold. I want to keep the blood going."

"I'll get the blood going."

He laughed. He looked at her and laughed again. She had real nerve. She said what she wanted and made him laugh. It always worked.

"Quit whipping yourself like that," he said. "You're getting red in the face."

"Is it 'cause I'm Catholic?"

"Is what 'cause you're Catholic?"

"That you don't like me."

"Who said I didn't like you?"

"Because I am Catholic, you know."

"I know you're Catholic."

"I mean really Catholic."

"I know you're really Catholic," he shouted. "Everybody knows you're really Catholic."

"What do you think of Billy Brent?"

"He got me into Boys' Chorus."

"They hate all of us, you and me, all of us."

Manfred looked alarmed.

"I'm Catholic," she explained. "You're a Jew. They hate both of us."

"They?"

"Us together. They hate us. The rest. The Protestants. It makes us . . ."

They were behind the shop building, tucked away, facing some backyards behind a long metal fence. The windows on the shop building were glazed. Shadows fell everywhere, tree shadows, building shadows, fence shadows. Their own, superimposed one on the other. "Anyway," she said, changing the subject again. She pulled out a crumpled pack of Old Golds. "Quick," she said. "Light me." He did what she said, burning his finger in the wind. Ignoring him, she puffed away to get the cigarette going, then hid it in the palm of her hand. "Ummmmm," she said, inhaling again. "You don't know what you're missing." Smoke blew in his face. "It makes your heart beat fast."

"That's not good for you."

"It makes you high in the head."

He watched her puffing away in silence, trying to shield her from view behind his own silhouette. If they were caught, he'd be out of Forest Park and at some other school within twenty-four hours. "My heart's beating fast," she said. "Right here." She grabbed his hand and thrust it inside her pea jacket. "Feel it?" There was the soft absorbent texture of cashmere, a saint's medallion on Laura's chest that pricked his finger, the whole strange surface of someone else's body. He moved his hand around, searching for Laura's heartbeat. Yes, he could feel it pumping on, accelerated by nicotine. He thought of his opium vial, resting in a desk drawer at home. That too made the heart beat faster, infinitely faster than nicotine. It made the head higher than Laura had ever dreamed possible. "Hurry," she said, sneaking another puff. He moved his hand around. A look of anxiety fretted Laura's mouth. She had perfectly round little breasts. "Feel it?" she asked. He cupped her left breast in his palm, that was where the heart was supposed to be, rubbed the nipple with his thumb, thinking of Hugo Bondi. This was not the way Hugo Bondi promised it would be. "It's really going like crazy," Manfred said.

"That's the spot," she answered. "Don't take your hand away."

"I can feel it pounding."

"Just move your thumb around." After a moment or two, she took a last puff on her cigarette, then ground it out under her heel. His thumb pressed on. "Oh, God," she said as her face suddenly contracted without warning. A mild shudder passed through her.

"What happened?" Manfred asked. "Did I hurt you?" He withdrew his hand, staring at it as though he expected to see spots of blood. "Are you all right?"

She brushed some ash off her lapel. She looked sleepy. "You're kind of sweet, you know," she said. "Sweet's the word, sweet and dumb." Slowly, they began to head back to the main building. There was hardly anybody out on the track today. It was too cold. "Do you really like Billy Brent?" she asked again, yawning.

"He's OK," Manfred answered cautiously.

"Told you about him and Adele, didn't I?"

"That's over with and I never believed it anyway."

"Oh, no, it's not."

"You're wrong," he said, glancing at her. "It's over. She told me. She tells me everything. She's got somebody else. She doesn't care about that stuff anymore. If it ever was true."

Laura brightened for the first time in their walk. "I'll show you, if you think you know so much."

"I don't want to see," he said.

"It's nothing," she said. "Only the alleyway."

"I don't want to see," he repeated. Now his own heart was beating hard.

"It's only the Kates's garage."

"I don't care."

"They've got a rug in there and a pillow."

He looked away.

"Well, don't say I didn't warn you," Laura said. "Just meet me there at five o'clock."

"I have Boys' Chorus."

"No, you don't. Boys' Chorus is in two days. I know your whole schedule, every hour of the day. I know all about you. You can't fool me. You meet me at the Kates's garage at five."

They were almost at the main building. She stopped, pretending to brush something off the front of his coat. "You're the best-looking boy I ever saw," she said for the second time. At her words, everything inside him began to run together, beyond his control. "You're better looking than Billy Brent. Even Adele thinks so."

He looked at his feet, his face burning.

"Five o'clock," she said. She moved on into the building. A Forester stood guard there. "The four-minute bell's rung," he said in a warning voice. "We *know*," Laura said. "Just move your butt," the Forester ordered. She still wore her concentrated look as she walked away, her sharp focused stare; but the pinched expression was gone, her face was clear and without strain. Manfred could hardly wait to get to the lavatory to take a look at his own face in the mirror. The best-looking boy she ever saw. Better looking than Billy Brent. Even Adele thought so.

Ah, Manfred . . .

By five-thirty, the sky had turned a shadowed purple at the horizon, a few early stars glittered overhead, and the kitchen lights on Fairfax Road were on in almost every house. Straight ahead, through the small window facing the alley, Manfred could see Mrs. Kates's ample silhouette moving back and forth in sudden bursts of activity, between the kitchen sink and stove, stove and Frigidaire, Frigidaire and sink. The kitchen was so small that you could stand in the middle of the floor and touch everything just by stretching out your arms. Mrs. Kates reached up for a pot, holding a butcher knife in one hand. The light caught the sloppy bun at the back of her head. Manfred could make out the huge tortoise-shell pin that was supposed to hold it all in place. Then, in a single energetic spasm, she disappeared again, into the front of the house. They were trespassing on Mrs. Kates's property. Mrs. Kates could put them all in jail if she only had the sense to check out her garage every now and then. A pot banged inside the house, she was back in the kitchen, there was the sound of running water, as clear as if it was running out of a spigot at Manfred's feet.

Beside him in the alleyway, Laura was stretching herself on her toes. One knee-high sock had fallen almost to her ankle. "Stand still," he breathed. "You're driving me crazy." "You don't have to whisper," she said. "They only have ears for themselves. Anyway this thing is made out of iron." Discreetly, she knuckled the wall of the garage. "It's tin," Manfred said, listening to the metal shiver faintly, as if it had been touched by the wind.

"Look," Laura said.

Manfred peered through the filthy window. He thought he could see a couple of moving figures inside the garage, vague, soft outlines, probably human, pacing the floor. A lighted candle was set on an upended box. In the yellow glow, hardly more than a gasp of fire, he could make out Billy Brent's curly hair. Adele was off to Billy's side, almost out of sight. The two figures kept circling the garage floor. One approached the other, then withdrew. Soon they stood face-to-face again, hardly moving. It seemed that both of them were talking at the same time, but their voices were inaudible. As they talked on, soundlessly, the Kates's cat began to creep after invisible prey in the alley, right behind Manfred. The bells on its collar gave off a faint sound, gone almost before it could be heard. Manfred hated the Kates's cat. Every time he came near it, he found a drop of blood in the palm of his hand or on the inside of his wrist. His own blood. The cat's tail rose in the air now, curling silently. The tip floated there as though weightless. In the dark, Manfred could make out the cat's anus. The dry ringed patch, without hair, was staring him in the face. There was a sudden clicking sound, the cat moved, the bells sounded, a can fell over. A tiny scream came from the Gordons' backyard next door. It was soon muffled, and a second later the alley was silent again.

"I hate cats," Manfred said. He began to grind his teeth. From the inside of the garage, he could hear a familiar mumble: Billy Brent.

"It's only a mouse or something," Laura said.

"Or something," Manfred repeated with a sarcastic edge.

"You're not looking. Pay attention now."

"Nothing's happening in there. I told you that before."

"Holy mother of God, forgive me, but are you dumb."

"This wasn't my idea, I didn't want to come here in the first place."

"But you came. Now get in here, close to me."

They stood alongside each other, staring in. Fighting his own resistance, Manfred focused shortsightedly on the dirty window. It was smeared with old grease marks around the edges and in the middle. In the dim light, if he concentrated hard enough, the marks formed recognizable patterns. He could make out East Prussia in the right-hand corner, where the glass was cracked, the jagged Italian boot, somewhat distorted, a little further down, to the west. He pretended to concentrate on the geography, cheek to cheek with Laura, looking for familiar terrain. Inside, they were still talking, but louder now. "I'm absolutely sure," Adele said. "So quit contradicting me. You're always giving me an argument, that's one thing I can depend on." That sounded like Adele to Manfred. It had her clarity and assurance. It contained her firmness, her strength. Billy didn't answer her. No words came out. Then, in the silence, the figures began to move around again. Billy's shadow went right up to the wall, along the garage floor, then broke in two as it began to climb in a long skinny line. Adele disappeared. Tracing the odd shapes of Europe on the windowpane, Manfred found Saxony and a hint of Scandinavia, dangling down. "Nothing's happening," Manfred said, turning away.

In the kitchen behind the garage, Mrs. Kates's arms moved in a circular motion. She was whipping something creamy. Another silhouette moved alongside her, peering over her shoulder. The second silhouette was proportioned just like the first, high-waisted and round from head to toe: Hilary Kates, pee-an-ist, with two fat braids of hair hanging down her back, like Heidi's. Around Manfred, all the nighttime lives were beginning on Fairfax Road. The women paced the kitchens like controlled whirlwinds, the children fingered their homework, testing the depths, the fathers were arriving home, down at the mouth, reaching for the *Sun* papers or the *News-Post* as soon as they walked in the door. After a day spent in town, fighting the world, it took them an hour or two before they were able to hold a conversation with their families. Down the street, in the dark house, Billy Brent's father was taking off his trolley conductor's uniform and hanging it in an upstairs closet. Every crease of the uniform was fastidiously set in place. Mr. Brent took better care of his conductor's uniform than of his own clothes. Opposite the Gordons', Jojo McAllister's old man, who was a failed lawyer, sat in the living room talking to himself. That was what he did every evening until supper was served. Jojo had told Manfred that he thought his father was going crazy. Talking to yourself in the dark was a sure sign. As for Laura's own poppa, he was on his way home from the Piscitelli countinghouse out on Cold Spring Lane, after toting up a bid on a new rowhouse project planned for the other side of town. If he got the contract, it would be another step towards getting the family out of Fairfax Road and over to Ten Hills, away from the Jews. Ten Hills was where Laura's poppa's heart was, it was where they all belonged, he believed. Ten Hills and the Piazza Venezia.

There was a sudden movement inside the garage. Billy Brent's shadow leaped to the ceiling. "I have to go now," Manfred said.

"Hold your horses," Laura said. "Something's happening. Get back here."

There was a slipping somewhere above them, a metallic scraping overhead, tiny bells sounded chaotically, then the Kates's cat came sliding down the roof headfirst, scrambling for a hold. It landed about five feet away, looking stunned, then, in an instant, put its nose in the air and twitched away. Its tail jerked ferociously. Its anus was aimed at them again. It carried itself high, as though no one were there, as though it were invisible. In a minute, it disappeared.

"I hate cats," Manfred said irritably.

Mrs. Kates was staring out of the kitchen window. Her face was set somberly, a hand shaded her eyes. "Don't move," Laura warned without moving her lips. "Stay where you are."

Inside the garage, the tiny flame threw a pink circle of light. Manfred could make out Billy Brent's curls again. Mrs. Kates peered out for a moment, then moved off, opening the Frigidaire door again. "Absolutely the last time," Manfred heard. It sounded like Adele's voice. "Self-respect."

He wished she hadn't said that. It sounded unnatural. It wasn't Adele's vocabulary. She never used pious words. He wished she'd just put on her coat and slip out of the door, without so much as a by-your-leave. Then quickly into the next yard, home, and Florence banging away in the kitchen and Max in the living room reading the paper back to front, and Florence's godawful meatballs heating on the stove; as though nothing had ever happened. He'd even welcome a little squabble at the dinner table if it meant having Adele back in the old ordinary way. Adele'd snap at him, he'd snap back, and Florence would lean over her plate, one hand dramatically covering her eyes, to hide her parental suffering. "That's enough," Max would say in conciliatory tones. "You're both old enough by now . . ."

Instead: "You think you rule the roost," Adele was saying.

"You talk too much," Billy interrupted. "You think you've got all the answers. Anybody ever tell you that?" He plucked at her arm.

"You're snagging my sweater," Adele said. "Don't do that. Don't do *that* either. Just leave me alone for once."

"You didn't have to come here today. Nobody forced you. You don't fool me. You came here of your own free will."

"Who's talking about free will? I've never tried to put anything over on you and you know it."

"What about that creep from Cleveland? Does he know it?"

"Oh, you're hopeless. You never remember a thing I tell you. It's Cincinnati."

"Does he know it?"

"Know what? I don't even remember what we're talking about. And don't touch me, I told you before. Don't come a step closer."

"Oh, Lord above," Laura murmured happily. "Can you see all right?"

"Nothing's happening in there. It's all talk. It's too cold out here." He was sweating.

In response, Laura ran her fingers lightly up the cleft between his buttocks. He twitched away. "Just cut it out," he said.

"You're really a sourball." She lifted her leg and pulled up her sock.

"I think it's disgusting."

"What's disgusting? Nothing's even happening. You said so yourself."

"Spying."

"Spying? You call this spying?"

"What do you call it?"

"Listen, I found this place for them. I arranged it. I helped furnish it. Those little flowers are mine. They would have had nowhere to go without me."

"It's still spying."

"Perfect Percy."

A car began to thread its way slowly down the alley, coming from the far end. The two headlights shone like giant cat's eyes.

"How did I ever get started with this thing anyway?" Adele was asking. "Somebody please tell me."

"What *thing*?"

"This thing. Us. You. Me."

"You know the answer better than I do," Billy Brent said. "You got started the way I did. You wanted it. You fell for it. You loved it. The It Girl." Billy laughed.

Adele chewed on her nails. "Save your jokes," she said. "You can stuff them in your pocket, they won't take up much room there. I mean why don't you just let go? There are a thousand girls out there better looking than I am. More loyal. Smarter, too. I mean, you want everything. Nothing's forever. It's time now. It's over. I like you, I want to keep on liking you."

Billy Brent pushed against her. "You think you're too good for me," he said. "You think I don't read books and shit like that."

"You know how I hate that kind of talk, Billy."

"Shit's shit."

"The broken record," she answered, turning her head away.

"I don't own a department store in Cleveland."

"Cincinnati, you jerk. I hate self-pity, too. And stop it, Billy, you're hurting me."

Manfred's eyes widened. He made a few growling sounds in his throat. Adele was beginning to cry. He had never seen her cry. Suddenly she looked like somebody else. She didn't look like Adele. She looked like her mother, mouth turned down bitterly at each corner, emotion draining from her eyes, anguish drilling her cheeks. Blurred like that through the dirty window, she looked about thirty years old.

"You bring out the worst in me, I swear. You always do." Billy's rosebud mouth pulled in. "For Christ's sake, stop crying Adele. Stop. Cunts," he said. "Always crying."

"Oh," Adele cried, hitting him hard.

"That's more like it," Laura said, looking around for confirmation. "What's this?" The headlights of a car were shining in her eyes from about twenty feet away. They seemed familiar. They were familiar. The lights came closer. They shone from Ettore Piscitelli's car, on its way home. "Holy mother of Christ," Laura shouted, "save me."

Manfred was already gone. Something had sucked the consciousness out of him. The giddiness swept him inside the yard to the door of the garage, a sense of powerful occasion drove him through. Suddenly facing Adele and Billy on the concrete floor, he was not prepared for the chilled air, for the knifelike cut of frost that came back at him right out of the tin walls. He pulled back a moment. The sad little candle burned dutifully. Artificial flowers were stuck on a ledge in an old vase. Both Adele and Billy had their shoes off. It made them look strangely vulnerable. Beneath their feet was a tattered forlorn carpet. Manfred thought he could see spiders everywhere. One hung over Adele's head in a luminescent web. Manfred thrust his right hand in the air, as though he could cut through the unhappiness with a single manly gesture. He kept it there for a moment, like a priest. "You have to stop this," he shouted at Billy Brent. His voice gurgled up and broke in two. A tin echo rattled back.

"Well, I'll be fucked," Billy Brent said mildly.

"I won't put up with it," Manfred shouted again.

"He won't put up with it. Now hear this. Did you hear this, Adele? He says he won't put up with it."

"I forbid it." He heard the absurd sound of his own words as he spoke. He was like Jojo McAllister's father. He was going crazy. He was talking to himself.

"You little jerk . . ." Adele said.

"Let me handle this," Billy said. He was half smiling. "Just stand aside. We don't want any broken bones in here."

"What do you think you're doing, Manfred . . ." she went on.

He stared at them both. There was a noise outside, a voice shouting in a strange language. Then a voice answering: "Poppa!" Manfred was freezing, he heard his teeth click together. He waited. It was two-to-one, and he was the one.

"Be a good boy now," Billy Brent said, "and answer your sister."

Manfred felt his back stiffen. His muscles tightened, but all he could do was chew on his upper lip.

"What's wrong, baby, cat got your tongue? I thought you were a big shot." Billy Brent began to beat his chest like Tarzan. "That's a big shot now," he said, pounding away. "See? That's a big shot."

"Quit acting like a fool," Adele said to Billy. "And how dare you follow me like this?" she cried at Manfred. "How dare you?" She shot a fist at him, catching him just above the elbow.

"Easy now," Billy Brent said.

By then they were no longer alone. Ettore Piscitelli was standing just inside the door, holding his daughter by the ear. He had stepped out of his car as Manfred entered the garage. He had left the headlights burning and the motor running. He was beside himself with righteousness. "So this is where you hang out?" he raged in a voice that seemed to shred like strips of metal. His eyes were hard as aggies. He held tight to Laura's ear. "Right under my nose, right under your mother's nose. You with your Jewboy and that dumbhead Swede. I know all about it, you don't have to tell me anything."

"Who are you calling a Swede?" Billy Brent asked, stepping forward.

"Get out of my way," Ettore Piscitelli said. He was about a foot shorter than Billy.

"Poppa," Laura cried out.

"Stand still, you, if you know what's good for you." He gave her ear a twist.

Mrs. Kates arrived then in her apron, rubbing her upper arms to keep warm. "Mr. Piscitelli?" she called out in a timid voice. Hilary peeped out from behind her mother's vast bottom, one braid falling over her shoulder. The look on her face said that history was being made tonight, right in her own backyard. Down below, the Kates's cat was rubbing against Mrs. Kates's ankles, tail in the air. "What are you all doing here in my husband's garage?" Mrs. Kates asked in a firmer voice. She stood up straight, attempting self-possession. "Where are your shoes, Adele? What are you doing running around my husband's garage without your shoes?" No one spoke. Mrs. Kates thrust Hilary behind her. "I told my husband months ago there was monkey business going on out here. When I saw the rug and the flowers, I knew something was fishy. What are you all doing here in my husband's garage?"

"Explain to the lady," Mr. Piscitelli said to Laura.

"Poppa."

"I'll explain," Adele said, reaching for her shoes.

"You know this is private property," Mrs. Kates went on. "You have no business in here."

"Yes, yes," Mr. Piscitelli interrupted.

"Then what's going on? I want to hear from the children," Mrs. Kates said. She was in control now. She knew it. The Law was on her side, she was Authority. Another moment passed. Her ordinary ample presence was already beginning to force them all back into themselves. She made them think of their own mothers, with shame and trembling. There was another moment or two of real confusion—overlapping voices, quiet anger, the babble of fear—then it was over. A kind of calm asserted itself. Mrs. Kates never did hear from the children. She retreated again into timidity. Laura hopped into the Piscitelli car alongside her father to make the rest of the trip home, six houses away. She was actually grinning, although in pain. Her father put the car into first and let it roll. After a moment, during which everyone forgot him, Billy Brent put on his shoes, slung his lumberjacket over his shoulder, and set off for his own dark house down the alley. He swaggered a little as he walked, but his stomach was contracting painfully at the evening's events. He had to sort it all out, he had to make something real he could hold onto out of the chaos. He always faced disaster that way, trying to keep his idea of himself untainted.

That left the three of them to face each other inside the tin garage. "Don't tell me," Mrs. Kates said as Adele opened her mouth to speak. "I don't want to hear." The candle blew out. It was a relief to them to be able to stand together in the dark, even though they could still see each other dimly. "Hilary? Where are you? I told you to go practice. And feed kitty. Where is she? She was just here, making up to me. Good heavens, the cold, we'll all get sick. Billy Brent. That's not hard to believe. But Adele Gordon? Don't tell me. And you?" she asked then, turning slowly to rake Manfred with a scouring look. "You, too? What does your mother have to say to all of this?"

His mother. But what could Mrs. Kates know about Manfred's mother? Mrs. Kates didn't wait for an answer. She had had enough, and what she didn't understand was not worth understanding. She turned her back on them (in her agitation, her sloppy bun was beginning to come loose) and headed for the kitchen. "Make sure the candle is out," she said over her shoulder. "The last thing I need is a fire out here. And take your belongings with you. All of them. And don't come back."

Adele and Manfred stared at each other. They had gotten off easy. Mrs. Kates looked as though she wanted to forget the whole thing. With luck, she would. Manfred began to move off, backing through the garage door. "Don't forget your shoes," he said to Adele in a stern voice. A flash of anxiety touched him as he watched her sit down on the old rug to pull on her saddles. She looked cowed and shrunken, sitting there bent over, intent on her laces. Her body seemed to have grown smaller, hunched in, it was just a quivering, diminished body, gone slack. He felt anxiety again, a sympathy, too, at the same time, that drew him to her. Her white face was like a fading moon, her eyes were cast down, her straight hair hung stringy over her cheeks. She kept brushing it away. How was it that he had never noticed how stringy her hair was? How pallid her skin? She was struggling with her shoes, like a child, grunting with the effort. Poor Adele. She was just like everybody else . . .

"Maybe," he said, feeling his jaw muscles pull back in resistance, "maybe you'll learn to stay in your own backyard from now on."

He despised the excruciating sense of moral superiority that forced those words out of him.

Chapter Five

While Manfred waited for the cruiser *Emden* to visit Baltimore on its round-the-world goodwill cruise, the Metropolitan Opera prepared for its own visit to the city, an annual conflux of social and musical hyperbole, which had provoked, some thought, near-hysteria over the years. Lester Ottinger agreed.

"You'd think three nights of opera was enough to transform a tank town into a cosmopolitan wonder." He spoke, over Friday night dinner with the Gordons, with his customary self-conscious drawl. "At least Cleveland gets eight performances and Philly has a night each week during the season. But us? Midnight'll strike and we'll still be bumpkins."

Clara turned aside at Lester's words. She had been concentrating on a thoughtful analysis of Richard Wagner's thematic progressions for Manfred's benefit, and she considered her husband's interruption as infra dig. Those were her own words. "Why do you think you have to talk as though you come from Cumberland or Hagerstown or someplace like that?" she asked. "I think you get a kick out of acting beneath your dignity. Infra dig suits you. You do it at the club, too. To prove what?"

He didn't know. All Lester knew was that he couldn't help himself. It had been going on for years. Whenever some of the old scarecrows approached his table at the club on Saturday night, something curdled in his blood and demanded instant public acknowledgment. "Well, if it ain't Fanny Mannheimer," he was likely to say to one of these bony apparitions (out of the side of his mouth, of course, like a yokel). "And I thought you was long gone." Or, in a single affected breath, "Shucks, cousin Bernice, you don't know your knee from your elbow, and even if you did, you'd insist on poking with your knee and kicking with your elbow." Then they would all laugh together, patronizingly, Fanny Mannheimer, cousin Bernice, and whoever else was around as if Lester was the village idiot, to be more or less fondly indulged;

and he would be able to return to his dinner, mission achieved, with the world still safely positioned at arm's length. But it exasperated Clara, it always had.

"In the opening forty-five minutes of the opera," Clara said, returning to the subject, "it is possible to hear virtually every leitmotiv that Wagner encrypted inside this magical score. Why, the prelude alone . . ." Here she lost her grip on the language, her emotion was so powerful. But it passed. "Have you heard *Tristan* in the opera house?" she went on, getting hold of herself. "Have you ever seen it actually *done*?"

"I saw *Rienzi* once," Manfred said in a monotone.

There was a pause. "I do not believe that *Rienzi* is in the Met's repertoire," Clara said. "I do not believe that we have ever heard *Rienzi*."

"I fell asleep three times," Manfred said.

"You fell asleep! But look here, Manfred, that won't do. It really won't. If you intend to use this priceless opportunity to catch up on your beauty rest, we'd be better off changing our plans."

"He's kidding you, Clara," Max said.

"Kidding me."

"He likes to make jokes," Florence said.

Across the table, Adele looked stony.

"I fell asleep at *Rienzi* three times," Manfred repeated. "Lots of people sleep at the opera."

"Not in our seats they don't," Clara said. "Your ticket is priceless. Flagstad, good Lord. And Melchior. And Ponselle. Surely, I've told you."

"You've told him," Lester said. "Twice tonight alone."

Still, Manfred knew that tickets were what she said they were: priceless. And he had one for each performance, formerly cousin Stewie's. It was the Met, after all, truly Grand. The Met was known even in Frankfurt. It was famous everywhere. He would do his duty by it, he would behave. He was all right as far as *Bohème* and *Carmen* were concerned. He knew them backwards and forwards, *Carmen* especially. He could do Don Jose's aria himself, he knew Escamillo's song, too. But when it came to *Tristan* there were problems. *Tristan* was four hours long. What would he do with himself for four hours? How would he get through the first two acts and still be awake for the Liebestod? All Manfred knew of *Tristan* was the Liebestod, from a recording that Oskar and Lotte let him play in Frankfurt. It went on and on, circling slowly around itself, like the moon circling the earth, throwing a baleful light on everything. The evening would be like an operatic Yom Kippur, time without end. Manfred hated Wagner. Richard Wagner oppressed him. Listening to his operas was like sitting in the middle of a blast furnace in which outsized human beings, wearing armorplate and helmets, screamed at each other about peculiar matters, like incest. Sitting at the Ottinger table, Manfred tried to show some interest. Somebody was now talking about the new baseball season. Somebody on the Orioles had been traded for somebody else. Somebody was always being traded for somebody else. Just last week, it had been Billy Brent for Bernard Potter. Opposite him, Adele fluttered her eyes helplessly. She, too, looked as if she didn't care. When she glanced his way, her half-closed eyes passed an inch above his forehead. When she spoke she directed her words to Uncle Lester at the head of the table. Even when Manfred knew that what she was saying was meant for him, she spoke to Uncle Lester. She didn't speak to Manfred on Fairfax Road either. He wasn't worth a greeting. When he came downstairs for breakfast, she got up from the table. When he knocked on the bathroom door, she flushed the toilet to cover the sound. He told himself that she despised him, that was what the scene in the garage had come down to. And it made him feel better to believe it. It relieved him of responsibility. I never loved her, he told himself, with satisfaction. Never. How could he have loved her? With that stringy hair, that dead white face. It was all a dream (not one of the good ones), and like any dream it had cast its little shadow for a moment or two and passed into oblivion; no one had even noticed. Good riddance then, as Hugo Bondi would say; good riddance to Adele Gordon. The truth was that the only people he really loved were Elsa and Julius Vogel and his brother, Kurt, Kurt perhaps most of all. He continued to send them postcards and letters while they waited together so patiently in Frankfurt for something to happen, cheery little messages that were stiff with lies about his own happiness. (He told as many lies as they did). He now signed his letters to Kurt "Your baby brother." Why not? That was what he was, what he had always been. It would throw Kurt off the trail. It would fool Elsa and Julius Vogel, and it was important to fool them all. At the table now, he slumped in his chair, and counting to ten under his breath, forced himself to think of something else. What had Mrs. Kates told them all about the garage? Not much, apparently. At least it seemed clear that she had gone easy on Manfred and Adele. Billy and Laura were the villains. Everybody had seen to that. Billy was Protestant, Laura Catholic, what could you expect? When Max had heard the news, whispered to him by Florence in her most ingratiating manner, there had been a brief excited outburst that lasted only a second or two; and that was all. Goyim, was what he was thinking, Billy and Laura. But it was another matter with the Piscitellis. The Piscitellis had the goods on their daughter at last. Laura was now a day-student at Mt. St. Agnes, an all-girl parochial school on the other side of town. Mt. St. Agnes took up most of Laura's time; that was one of the things the Piscitellis were paying for. She hadn't been seen on Fairfax Road in a week, and Manfred felt sorry for her. "You're the best-looking boy I ever saw," she had said more than once, making a friend for life.

"Baseball is very tedious," Clara was saying at the table.

"Max loves baseball," Florence said. Since her pregnancy, the words no longer sizzled in Florence's mouth. These days, she was much calmer, she was even benign, swelling patiently while the future lay in wait inside her.

"We're minor league in baseball, too." Lester put in, fingering his glass of bourbon.

"You want baseball, you want opera, go out and get 'em," Clara responded.

"Yes, ma'am," Lester said, acknowledging his wife's command with a courtly nod. "Now, is there any chance of bringing this Sabbath collation to a conclusion?"

"Just tinkle the bell," Clara said. "How can Shorty know you want him unless you tinkle the bell?"

"I will not tinkle the bell. You know how I feel about tinkling the bell."

"It's too late in the day for quixotic behavior."

"I won't ring a bell to summon another human being. It's demeaning."

"Oh, the sensibilities of men."

"If you're so tough, you ring."

And she did, after Manfred lazily passed her the little silver bell. Crazy, he thought. It was so simple. They were all crazy in America.

Quarterly exams were coming up. They came up four times a year, like bitter little thorns with a regular season of their own. A lot of students were worried, as usual; even worse in some cases. That was the intent of quarterlies, to surface responsible concern among the student body and concentrate the mental faculties. Jojo McAllister had already panicked. He lived his days in a state of terror and despair. The skin around his cuticles was all bloody, the prospect of failure was already an open wound. Down the street, Billy Brent now gave an extra half-hour each afternoon to his homework. He supplemented that with an additional half-hour at ten o'clock, before he went to bed. He had it all worked out. Cramming was Billy's specialty. Adele, on the other hand, had it all at her fingertips. Her notebooks might be a mess, filled with drawings and cartoons and compelling messages scrawled to herself, but they also contained everything that mattered as far as quarterlies were concerned. Notes, facts, references, all the right ones. Besides, her memory retained everything. Like Billy, she knew what she was doing; she just did it another way.

These were Manfred's first quarterlies. The word had already passed into the realm of pure sound for him, he had heard it said aloud so many times in recent weeks. Naturally, he thought, he would fail everything, like Jojo. The two of them would stay in the tenth grade forever. He began to review American History while he brushed his teeth. He recited theorems of plane geometry in the shower. He practiced irregular French verbs as he sat on the toilet. A wart suddenly appeared on the middle finger of his right hand. There was a blaze of eczema in the crook of both elbows. He even managed to vomit from nerves after one of Florence's dinners.

Sometimes in the morning now, he could hardly move when he awoke. He lay in bed and stared at the ceiling, almost paralyzed by inertia. Slowly turning his head from left to right, as though it hung on a rusty hinge, he thought he could smell Adele's new perfume, bought since Christmas. Like them all, Adele had her own unchanging clinging smells, carried with her wherever she went; mysteriously, they could come right through the wall sometimes, especially at night and in the morning, stopping Manfred in his tracks, overcoming his will. She was moving sluggishly around her room now. The sound transfixed him. It stopped and started; it hesitated and groped; it was slow and lazy. He could see her ridiculous bunny slippers, with the white fur peeking out. She was probably looking for them now. Her tiny pink panties, which you could tell were weightless, hung crookedly from the arm of her desk chair. Her brassiere, with its little pouches for her nipples, swung on the doorknob. Sometimes he could see the outline of her nipples through her sweater, puckered out symmetrically. She didn't seem to mind. It was a mess in Adele's bedroom. It was like Adele's notebooks, absolutely incoherent to the rest of the world. He heard one slipper drop, then the other. Then it was silent. She was poky, all right. She was lazy. They were both lazy. Lying in bed every morning, staring at a spreading crack in the ceiling, he liked to stroke his thighs and listen to himself begin to purr with quiet pleasure, while Adele slowly came to life next door.

"This is not the assigned subject," Athol Burns said, frowning, when Manfred handed in his first weekly essay.

"It isn't?" Manfred asked, somehow not at all surprised.

"I told you to write it down," Athol Burns said. "Sure as I'm sitting here, I told you to write it down. Writing maketh the exact man. Didn't I say that? Do you believe me now? You're going to have to get it right, otherwise there's no point to the struggle. Getting it right is the essential thing. Get it right and you're home free. Get it right and you're in the clear. Think, boy, think. I don't mean brood, I don't mean dream, I mean think. Use those precious little cells"—here he tapped his temple with his forefinger—"that God gave you, gave every one of us. They're the real mark of our humanity, the stamp of our uniqueness among so much other uniqueness. Do you follow? Are you with me? Don't let the dark forces get at you, boy. Escape them. Throw them off. They'll make mischief, given half a chance. That's their job in life, to bring you down. Throttle the dark forces, erase them, obliterate the black spirit. Use your brain. That's what it's there for. Use your brain and it will never fail you. Do you feel the powers of darkness? Have you a sense of the irrational, of wrongdoing, of error and sin? Now don't make the mistake of thinking that I'm talking about religion. I'm not talking about religion. Religion has no place here. Religion and state—" here he pretended to slit his throat—"*kaput*. The Founding Fathers saw to that. They knew what they were doing. No, I refer to ethics. Man's noblest venture. E-t-h-i-c-s. His highest flight. His most rapturous dream. What makes Man man, worthy of God's attention. Do you catch my meaning? Do you have a clue? Do you?"

"Yes."

"Yes what?"

"Yes, I have a clue."

"Boy?"

"Sir?"

"Now you're getting it. You just used your brain."

"Yes, sir."

"Are we making progress?"

"We are making progress."

"Are we using our heads?"

"Using our heads."

"Boy?"

"Sir?"

"Are we ready for quarterlies?"

"Ready."

"No more fuck books?"

"No more fuck books." He snapped the word out. He heard his teeth click.

"We are on our way. Join me in thanksgiving."

"Sir."

"Say a hallelujah."

"Sir?"

"Next essay."

"Yes."

"'The Meaning of Ethics.'"

"'Meaning of Ethics.'"

"For God's sake, boy, write it down."

"Yes, sir."

"And don't go looking it up in Webster's. I want you to find the real meaning, on your own."

"The real meaning," Manfred said, writing away.

"Three hundred words."

"Yes, sir."

"There's the bell."

They both stopped to listen.

"On your way now."

"On my way." He gathered his books. He stood up. Sweat trickled down his rib cage. He moved towards the door. Athol Burns sat behind him, fingers pyramided in front of his nose. He seemed deep in passionate thought.

His coppery skin looked purple. A pink vein ran in a corner of his forehead. Mr. Burns's eyes swept the carpet below him, like a vacuum cleaner. He looked as though he was praying. Athol Burns was crazy, Manfred was thinking. It was so simple. He was crazy. He was as crazy as the rest of them.

"Listen," Max said to the figure standing in the doorway. "I've been meaning to talk to you." In his easy chair, he rustled the newspaper in front of him.

"To me?"

"Come on, Manfred. You've been avoiding me long enough."

Manfred came into the living room.

"Don't stand there like that. Have a seat."

"I have a lot of homework to do. It's quarterlies."

"I'm sure it can wait for once." Max put the paper down. "Don't stand way over there," he said. "Come closer."

Manfred moved in.

"Listen, I don't want to butt in. I don't like butting in. I don't like it at all. But what's all this about anyway?"

"What's what all about?"

"The Kates's garage. You. Adele." Max ran his fingers through his beautiful hair.

"Oh, the Kates's garage. You'll have to ask Adele. I don't know about the Kates's garage. Adele can tell you. She knows all about it."

"Sit down, I said. Don't fidget so. I'm not finished. I'm worried about you."

Manfred made a face.

"I mean it. Something's going wrong. You don't have to be a genius to see that. Are you happy with us? Should I be worrying about that? I mean, am I coming down too hard on you? Am I doing what I should be doing? You have to tell me. I'm not clairvoyant."

"You're not too hard on me."

"You don't sound full of conviction. You don't even sound interested. That's part of what I mean. Is it me, is it my wife or my daughter?"

"Everything's OK."

"Call me Dad. For once. Come on, try it, it won't hurt you."

There was a silence.

"You see?"

"I said everything's OK."

"Sure. Everything's OK. I didn't say it wasn't."

"Can I go now?"

"Sit down a minute. Look,"—waiting a moment for Manfred—"I've doubled my income since you got here. Florence has doubled *herself*. Ever think of that?" Max smiled weakly from his velvet chair. "You're our talisman. It's clear. You bring us good fortune. Before you came . . . well."

Manfred pursed his lips appreciatively.

"Looking forward to the baby?"

"Everybody is."

"I'm talking about you."

"Sure."

"It'll be nice, having a new baby around."

"Sure." Manfred began to fool with his schoolbooks, flipping the pages of Carl Schurz's *Memoirs*.

"Maybe you'll work at the store this summer," Max said. "God knows there's plenty to do down there, it's growing so nicely. Maybe you'll be our advertising contact. You could handle that. You're smart enough."

"Summer is vacation."

"I'm talking part-time."

"Sure. Part-time."

"OK." Max picked up the paper, put it down again. "The Piscitellis are moving," he said. "The For Sale sign is up."

"I saw it."

"It sure stirred up a mess."

"They've been planning to move for a long time."

"What actually happened out there? Nobody seems to have it really clear."

"Out there?"

"In the alley."

"Ask Adele, she'll tell you."

"Ask Adele. Easier said."

"Adele knows. She'll tell you I'm telling the truth."

"You're not telling me anything. Neither is Mrs. Kates."

"There's nothing to tell. It's the truth."

"Have it your way," Max said after a pause. "Maybe you'll go to Atlantic

City this summer with Uncle Lester and Aunt Clara. For a couple of weeks." Max crossed his feet on the ottoman. "Or maybe I could talk old man Hagenzack into giving you a special membership. Would you like that?"

"Sure."

"You've been doing pretty well in school."

Manfred nodded politely.

"Probation aside," Max added. "Tell me about it, about probation. Tell me what happened in school."

"Well," Manfred said, taking an extra breath. "Miss Feinstein thinks I'm a genius, it's because I'm Jewish, and Miss Klinesmith lets me teach her class. In French. Mr. Lambrino doesn't like me, it's because of Jojo McAllister, but it doesn't make any difference, I get good marks anyway."

"Jojo McAllister?"

"He lives across the street. You know who I mean."

"You mean the kid with the jug ears?"

"I guess so."

"He's your friend? I thought he was a dumbbell."

"I help him."

"Not to your own detriment, I hope."

"What does that word mean?"

"Forget it," Max said, waving it away. "I didn't really mean what I was saying. It was a reflex, a defensive one, the lion protecting his cub. You know, you won't believe this, but you'll forget all this eventually."

Manfred looked blank.

"You will."

"No." Manfred shook his head.

"Oh, yes, you will, I know what I'm talking about. You'll forget all this because it will be unbearable to remember. It happens to everyone. Selective amnesia about the past. It's the only way the world keeps going. You'll see."

"I don't think so."

"You take my word for it. I know from happy experience. We think we can remember everything in detail, but all it is is the shadow of the past hanging over us, the shadow of a shadow that won't go away, that's all. And it's enough. It's more than enough. If we really remembered our lives, if we had that ability, if we could relive it all in memory, we'd just cave in, all of us, we'd get the bends every morning."

Manfred decided that Max was telling him that his life would be as though it had never happened; that it was worthless.

"I don't think you're very interested in what I'm saying. I can't tell when you're interested in anything anymore."

"Are we finished yet?"

Max paused. "Is there anything else you want to talk about?" he asked.

"That's all."

"All right." Max picked up the paper. "If you have anything more to say, now's the time to say it, you know."

"I don't think so."

"Don't be such a stranger."

"No."

"And I promise not to butt in. I'm not nosey."

The room closed in on itself then, as though it were folding up like a cardboard house in a child's pop-up book. The corners came together, neatly, the floor rose in the air, the ceiling sank in. Max opened the paper and began to read the editorials in the usual way, while Manfred headed for the stairway, books under his arm. He was reciting Euclid under his breath and reaching for his wallet with his free hand. He wanted to have a look at Kurt's picture. He had to see a Vogel face in front of him. He couldn't wait. He was overcome by the need several times a day now, in school and elsewhere, and he had already learned that he could not resist it.

Chapter Six

On Fairfax Road, Lester's horn blew twice. "They're here," Florence said from the window. She turned to Manfred. "Straighten your tie," she ordered, looking him up and down. "Fix the clasp now." But she needn't have spoken. He had on Aunt Jenny's going-away present, the one with the polo player and the horse, and it was already perfect, with a tiny, hard, fastidious knot pressing against his Adam's apple. Manfred was also wearing a blue serge suit with long pants, a white shirt, and new American shoes bought at the Hub downtown. The shoes had flaring wing tips, which he admired, the soles had a slick, wooden sheen. They were still untouched. Everything he had on was untouched, except his underwear. Max put the paper down, peered out at him. "Not bad," he commented, taking off his reading glasses to see better. The horn blew again. "Oh, shut up," Adele muttered. She turned to face Manfred. "You look pretty snazzy," she said, gazing at the wall behind him. He could barely hear her. The sentence seemed to stick in her throat. They were the first words she had spoken to Manfred in nineteen days, since the episode in the garage. Was it peace at last? Her relentless silence at the beginning, which had lasted for a week or so, was followed by a long series of punishing sounds—grunts, moans, indifferent sighs, wet gurgles—meant mainly for his ears, then these four almost inaudible words offered tonight: "You look pretty snazzy." It could mean that she was no longer furious at him. Perhaps, living side by side, they would now become friends. That was all he wanted; anything else would be false; the old Adele was gone.

He was winding Max's white silk scarf around his neck. It was just like Kurt's, with the same high gleam. Max had pressed it on him. "If I wear it twice a year, it's a lot," Max said. "It's a waste for you not to wear it tonight." Without acknowledging Adele's compliment, Manfred let the cool fringes of the scarf slip through his fingers. They were like icy sand. "Hold your horses," Florence shouted from the front door at another blast. "Do you have everything?" Max asked. Manfred patted his wallet and his handkerchief. He put on his coat. "Enjoy yourself," Florence cried. She clasped her hands in front of her mouth to conceal the pleasure.

"Hey," Max called as Manfred went out the door. "You didn't even say good night." Manfred ran down the porch steps. "He didn't even say good night," Max said. "Bye," Adele called, watching him go. Florence pressed her face to a pane of glass. Without turning around, Manfred gave a halfhearted wave. He checked his fly buttons. He was glad to be out of the house.

It was Lester's new convertible sitting at the curb. "Oh, no," Manfred said when he saw it. After Clara got out to give him room to squeeze into the crawl space, he hoisted himself down crosswise onto his left haunch. He was an expert at it by now. "I put a towel down," Clara told him from the sidewalk. "You don't have to worry about getting yourself dirty. Why we couldn't take the other car is beyond me. It's ridiculous." Lester sat behind the wheel, fingering his black tie. "For heaven's sake, quit yakking and get in," he said to his wife. "We're already late." The door slammed. Lester released the emergency brake. "You all right back there?" he asked, checking Manfred in the rearview mirror. Lester felt for his tie again. The motor turned over with a heady flush and off they roared. "I don't want to be late," Lester shouted in an excited voice, over the engine.

Lester too was wearing a white silk scarf, under an overcoat with a nappy velvet collar. He held himself stiffly as they crossed the street, after parking the car, as though all he could think about was how he looked. So did Manfred, feeling the newness of his suit and his wing tip shoes. "Watch the traffic now," Lester said, gazing right and left. "There go the Wertheims," Clara said over her shoulder. Clara's long dress—visible in front of them—was spotted with huge coral blossoms. The rest of it was black. Over the dress, Clara wore a boxy mink jacket. When they reached the curb, all three had to jump aside to avoid being hit by a limousine. "It's worth a man's life tonight," Lester said, taking his wife's elbow decisively.

"Don't go losing us in this crowd," he warned Manfred a moment later. There was no need to worry. Manfred was stuck to his side. Just by turning his head a little, he could see the bright moon of Lester's balding head shining under the marquee lights. Everything about Lester, in fact, was shining, down to his black patent shoes. Manfred could also see, above his silk scarf, that Lester had just had a haircut, probably that very afternoon. It had left his soft white neck totally exposed, as vulnerable as a child's, and the prickly line where the clippers had done their job was clearly visible. "Glad to be here," Lester acknowledged, as someone greeted him with a shout. He began to thread his way through the crowd, shouting his own greetings right and left. On the other side of him, Clara was saying something to an old lady wearing a white fur jacket and a tiara glittering with zircons and a single real diamond. The old lady had her right ear cupped in her hand. "Dear Mrs. Binswanger," Clara was saying. "We all adore Bohème. The whole world loves Puccini. Can there be a doubt?" She had to ask the question twice, in a loud patient voice, while the old lady strained to hear what she was saying. "Come along," Lester called, trying to marshal them in front of him. "We don't want to miss curtain." Aunt Clara had black mascara on her eyes tonight and the mascara had already spotted her cheeks. Driving downtown in the car, Lester had warned her about just such a possibility. "Of course you look positively smashing, my dear," he said, "but it's not as though you're exactly an expert at that kind of thing." Now, as she bent down to poke Mrs. Binswanger in the chest with her forefinger, there were also green smudges, like birds' feet, beginning to appear over her eyelids. "Please," Lester said to them. "Let's get a move on." He marshalled them again, tugging at Clara's sleeve. Inside the lobby finally, where it was almost impossible to move for the crowd, he had them wait in a corner while he checked Manfred's coat with his own, and finally, after a stop in the men's room—"semper paratus," Lester said to Manfred, standing in line at the urinal—the Ottingers and Manfred proceeded with overheated dignity into the auditorium, where white-gloved ushers, stationed nervously at the head of the aisles, waited to seat the audience.

The Lyric turned out to be a huge austere rectangle, shaped like a coffin, with none of the soft womblike curves that Manfred remembered from Frankfurt's baroque opera house. A right-angled balcony ran around three sides of the auditorium. On the walls behind the balcony and over the proscenium were engraved the names of music's immortals. Mozart, Beethoven, Gounod, Gluck, and their companions were chiseled in Puritan glory, one after the other. "Ah, Mr. Ottinger," one of the ushers said. "Nice to see you again, sir." In the boxes lining each side of the orchestra floor, wicker chairs were set out as though a garden party were about to take place. "Amalie, my sweet," Clara whispered, leaning forward for a kiss. While she introduced Manfred—"Mrs. Gruenberg," she said in a hard mannered voice, "Mrs. Kempner, Mrs. Strousse"—he kept biting his lips. He also had a crick in his neck from sitting in the back of the convertible. Clara kissed everyone self-consciously. Even Uncle Lester was kissing someone.

"Rudolfo," Manfred read in the program a few minutes later. He was seated between Lester and Clara in the middle of the orchestra. "A struggling painter in his garret on the Left Bank . . ." He knew all about that, about Rudolfo and Mimi and Musetta and all the rest, but he read it through to the end anyway. In the program, it sounded as though a lot of things were going to happen during the evening, when actually, he knew, it could all be over in a half-hour if they just left out the music. It was always like that with opera. Clara leaned forward to speak to the Haases, who were sitting in the row in front of them. She made one of her affected remarks, laughing in a peculiar false way, and Mrs. Haas replied in kind. "Schaunard," Manfred read, burying his head in the program at the sound of their voices. "Marcello. Colline."

"Can you see all right?" Lester asked Manfred. There was a tiny scab on Lester's chin where he had cut himself shaving. He kept clearing his throat nervously, as though he was warming up to sing a leading part himself, and in his right hand, he held his program rolled up like a weapon. One knee jiggled up and down.

Manfred heard Clara's voice again, over all the other voices. The Haases turned around to make a joke. Lester threw back his head and laughed while Manfred eyed him critically. Lester was laughing entirely too much tonight, when a smile would do just as well. He sounded as affected as Clara. He wouldn't stop jiggling his knee, he wouldn't sit still. Across the floor, there was a sudden flurry in one of the boxes. Two policemen began to rearrange the wicker chairs. "The governor," Lester said, stiff again. He put his rolled-up program to his eye and looked through it as though it were a spyglass. "And if I'm not mistaken, there's the Soviet ambassador," he added, swinging the program a few degrees to the right. A flute noise spiraled from the orchestra pit. Strings began to tune. The drummer gave a single whack on his bass drum, exciting everybody. Governor Nice and his wife entered their box while the audience on their side of the theater applauded. The governor was very fat. "Senator Tydings," Lester said, pointing to another box. In a moment, there was a great sighing sound in the hall, almost in one voice, the lights went down, flickered a second or two, then went off for good. Mrs. Haas slipped a cough drop into her mouth and turned around to offer the box to Clara. The conductor sidled in modestly, nodded to the players, then bowed to the audience. There was a splatter of applause, spreading front to back like light rain. A shiver went through the house as four jarring notes sounded from the orchestra, Manfred swallowed some air, and the curtain was up.

At ten minutes past eleven, when the opera was finally over, an antique jewel-box was presented onstage to Lucrezia Bori. Lucrezia Bori, who was the evening's Mimi, was about to retire from the Metropolitan Opera, and the jewel-box was Baltimore's reward for long and honorable service. The jewel-box had belonged to an Empress of France, a gentleman in white tie and tails explained to the audience as he offered it to Madame Bori. And it was only fitting, he added, while everyone clapped, that Madame Bori, as the finest jewel of them all, should possess it. Manfred agreed. He could see that everything about her was honed to perfection. It had been that way all evening. She was fragile, small-boned and white-skinned, with two carefully arched lines drawn in black where her eyebrows had once been. She moved slowly, with crisp, tiny gestures that could be seen from every seat. Everything she did was tiny and clear. Manfred even liked the way her voice had a tiny edge of acid to it. While she made her acknowledgment speech, he remembered the tenor's final renunciation. "Mee-mee," he had screamed when she died onstage. The tenor was Nino Martini, a movie star. Nino Martini was prettier than Lucrezia Bori, but he didn't sing as well. Nevertheless, his screams shook Manfred. When the orchestra began to sweep along beneath them, like water rushing in at high tide, he was doubly shaken. The tears started in his eyes. Blood filled his head. Sitting there, trying to hide in his seat, he almost forgot that that was what they had all come for.

Now a great affectionate burst of laughter came from the audience. Madame Bori had said something funny in her charming Spanish accent. She was bowing gently, acknowledging them all with a perfect gesture of her right hand. She held the jewel-box up for everyone to see. "Brava," someone shouted. Others joined in. There was a tremendous racket of foot-stamping then. They were all still applauding when the houselights came on and it was time to go.

"Audience of 4102, including 1104 standees, cheer Met Bohème," the *Sun* said the next morning. It was the headline of a story that was only one of many filling a whole page in the paper. The page also included pictures of three debutantes, Nancy Symington (in a Japanese coat), Grace Koppelman (in white fur, the season's rage), and Hanna Wright (holding an unexplained bouquet of roses); a feature story on the logistics of touring the Met; a rundown of the evening's parties preceding the performance; a critical review (favorable but not gushing); and an extended list of season patrons. Governor and Mrs. Harry Nice, Manfred read. Signor Augusto Rosso, Royal Italian Ambassador and party. Ambassador Alexander Antonovich Troyanovsky, USSR, and party. James Clement Dunn, special assistant to the Secretary of State. And then, picking out the Jews without even being aware of it: Miss Sarah Westheimer. Miss Naomi Hendler. Mrs. Max Hochschild. Mr. & Mrs. Jacob Haas. Mrs. Henry Binswanger. Mrs. Moses Ottenheimer. Mr. & Mrs. Lester Ottinger, Mr. and Mrs. Sydney Wertheim, Miss Rose Hamburger, Mr. Manfred Vogel . . .

"Well, how was it?" Florence asked at breakfast. She was sipping her coffee and trying not to sound too forward. Over her nightgown, she wore an ankle-length housedress, torn at the hem.

"It?" He was reading his name for the third time.

"Come on, Manfred," Florence said. "I'm dying to know."

Manfred could hear Adele rustling around upstairs. She was taking her time, as usual, poking around. These days she spent fifteen minutes each morning brushing out her hair. He still listened for her every morning, he was still intent, but he knew it was only habit, a mere echo of old feeling. And good riddance, he said to himself again, good riddance. Her father was on his way down now. Manfred could hear him coming down flat-footed on each tread. He would arrive at the breakfast table with a sober face made even soberer by the prospect of the morning's news from Europe and especially the day's problems at M. Gordon & Son. With success, Max suddenly had a lot on his mind. The sober look was something new. In the morning now, over breakfast every day, responsibility hung on him like wet moss.

"You're not eating my oatmeal," Florence said.

"I'm full."

"Tell me about the opera."

Manfred thrust the paper at her. "It's all there," he said. "Whoa," she cried, spilling her coffee. He sat back in his chair while she struggled to put the paper in order. She could read all about the opera in the *Sun*, like everybody else. There were thousands of words on the subject. There were pictures, too. It was better that way, he wouldn't have to brag to show off. There were no Gordons in the stories in the *Sun*. There never would be. Grace Koppelman didn't wear a torn housedress over her nightgown, you could be sure of that. Hanna Wright didn't have a faint cloud of moustache hovering cruelly over each corner of her mouth. No jewel-box would ever be given or received in that house, however rich and successful they all became. There was nothing in fact on Fairfax Road that had anything to do with the opera, Manfred knew, eyeing Max as he entered the breakfast room, looking ridiculously self-important, except Manfred Vogel himself.

That evening, *Tristan* was even longer than he had feared, four hours and considerably more, a slow languishing tempo oozing into the night and beyond. It was long for everyone and enthralling for some. Mrs. Haas lightly snored through Isolde's narrative in the first act, a cough drop slowly melting between her tongue and her cheek. Behind her, Lester sat as stiff as the night before, every now and then shifting in his seat from one buttock to the other, letting out a little moan under his breath. He had another tiny scab on his chin tonight, just below the first. Manfred heard vague sighs, sniffed overtones of bourbon. On his right, Clara sat at attention for the occasion, an attention not to be blunted by languor or slow tempos. She was dressed in black again, with a white lace collar and prim lace cuffs on her sleeves. She wore no makeup tonight. Her face was as bare as a stone. It was her severe German side, deliberately chosen to contrast with the bright coral flowers worn in honor of Puccini last night. Her nostrils flared with the music. Her eyelids fluttered. She looked like a great Wagnerian soprano herself. In preparation for the evening, she hadn't smiled once since the Ottingers had picked up Manfred on Fairfax Road.

"Give 'em hell," Max had called when Manfred went out the door.

"Bye," Adele had said again, in a small voice, watching him go as she had the night before.

"A butterfly," Florence said from the window. "He thinks he's a butterfly." Her smile turned down happily, like an inverted half-moon.

"Love potion," Clara whispered ominously during the prelude. "Longing for Tristan," she added, a few minutes later. After a while he began to understand. She was signalling the leitmotivs on which she had discoursed all the way from the suburbs to Mt. Royal Avenue, where Lester had parked the convertible again across the street from the Lyric. She knew everything about Tristan. She knew everything about Richard Wagner. She wanted to share it all with him, giving him something to concentrate on for the evening, making it all even bigger than it actually was. In profile, when Manfred turned to glance at her, he thought he could almost see Max's face thrust forward at the stage. Sister and brother had the same pious, grave face, heavy-boned and responsible, promising safety. But he had to admit that Max was handsomer than Clara. The strong bones were kinder to a man. Looking at their somber features, it was hard for Manfred to believe that they were both capable of facetious joking, Max especially. "Sea," she hissed. Richard Wagner was very clever. It was like unravelling a tapestry. "Love potion," Clara said again, poking Manfred in the arm. That meant drugs, he knew, which Tristan and Isolde were going to share. The theater was very still, the matter was serious. When the curtain had gone up, Manfred could hardly see what was happening. There was some confused action, then, on the stage, two women seemed to be embalmed in a deathlike fog. Minutes passed, a half-hour. Stately figures moved about slowly, declaiming. Mrs. Haas was primping her hair. Lester coughed an alcoholic fume. Clara said something that got lost in the voices from the stage.

Governor Nice and his wife had not shown up tonight. (It must be the press of state business, Lester had said, laughing before the curtain went up.) Ambassador Troyanovsky was there again, with his party. So was Signor Rosso, with his. Other gentlemen and their guests filled the boxes behind them. The women all wore corsages. "My God," Clara breathed. It was the sound of Kirsten Flagstad's voice, suddenly coming at them like a pillar of fire, and Mr. Melchior was almost her match. Sailors shouted on stage and began to sing in loud burly harmonies. They were on a ship, sailing west. They were either furling or unfurling the sails. A backdrop glowed in the evening sun. Manfred could see that it was torn at one corner. The audience stirred. Mr. Melchior sounded like a troop of well-schooled buglers. There were a lot of leitmotivs going past now, too many for Clara to continue her catalogue. Love potion, Manfred said to himself. Sea. They all settled back in their seats, grateful for the reviving sound of the men's chorus. It was very impressive. He wished that Miss Tubman and her Boys could hear it. *Tristan* was very different from Puccini. There were hardly any resemblances. Not only did it go on and on, but it was outside any world he had ever known, enlarged beyond normal boundaries. You could hardly touch it, it expanded so far. Music and words flew into space. Even the boredom, when it came, spread treacherously beyond normal boundaries.

"Kurvenal's song of mockery," Clara said, sometime later, in a voice that quavered. Her intoxicated passion paralleled the action on stage. Onstage, the lovers were doped by love. They could hardly move. Destiny was sealed. The seconds passed like hours. Clara had claimed that there were forty-two leitmotivs in the whole of *Tristan and Isolde*. "Forty-one," Lester had corrected her before the performance. He knew as much about Richard Wagner as she did, but it was clear that he could take it or leave it, the way he could take or leave everything. No such fusion of music and drama had ever been known in the history of the theater, Clara had said, ignoring her husband. At her words, Manfred nodded knowingly.

"Spiritual excitement," Clara now said in her seat at the Lyric, while Manfred thought of Fairfax Road and other places. The season was beginning to change again. The air was warm outside. Spring was the same everywhere. He had seen a fresh bud thrusting up on the hedge in front of the Piscitelli house. Behind it stood the For Sale sign. Up the street, where Billy Brent lived, a lone forsythia strained towards the sun. Billy Brent was busy these days worrying about quarterlies, like everyone else. If he didn't get a scholarship he wouldn't go off to Tidewater. Everybody knew he was desperate to go off to Tidewater. Everybody knew that once he went off, once he left the dark house on Fairfax Road, he would never come back. And where would Hugo Bondi go when he set off for the second time? And the Heller Brothers or Erwin Marks? Where would they go, on the run from the old realities? But that was another life. It had happened long before the story of Tristan. There was even more that had happened before that, countless events, another family, another city in another place, everything chained together by history. Oh, Lord, Manfred thought, feeling his forehead flush and grow heated. His eyes closed, his head fell sharply forward.

When he awoke, Tristan was singing to himself. *"Muss ich wachen?"* Manfred heard, as he opened his eyes. That was a pretty good joke. Lauritz Melchior didn't sound sleepy at all. What had he missed? The two lovers held each other. Together they filled half the stage. A kind of molten sound poured over the blue lights at the foot of the proscenium. Manfred had a memory of Tristan and Isolde singing together in his sleep. He had heard it all just below the level of consciousness, the whole rapturous drugged exchange, the pillar of fire and the bugle troop entwined. The orchestra shimmered invisibly. It was trying to play beyond itself, beyond the boundaries. Manfred could see the conductor's hands sweep the air. "Spiritual exaltation," Clara said again, hardly moving. *"Muss ich wachen?"* He was awake. He was at full attention. *"O ewige nacht."* Oh, endless night. Sitting there half-paralyzed, Manfred had to remind himself that Tristan and Isolde were singing in German. They had been singing in German all night, and he had barely noticed. It was so strange, hearing the heavy words coming over the footlights, the stone-hard consonants and exaggerated syllables plodding their way, one by one, in his direction. All night, squeezed in his seat between the dreaming bodies of his Aunt and Uncle, he had been listening to this siren voice of his childhood. It was like a call to arms. *"Ewig,"* they sang. *"Endlos."* He had to listen carefully to be sure that he heard right. Then it was over at last. A wall of applause rose in the air. Cheers followed. The curtain was down. Bravo, he screamed, startled to hear his own voice shrilling with all the others. His cheeks were burning. He forgot everything else. Tonight there were no speeches after the performance.

He had to miss *Carmen* the next night. It was Rosa Ponselle's *Carmen*, it was the opera he loved the best, but he was "under the weather," as Florence put it. He barely got through school that day. He had to skip his coaching session with Jojo McAllister. He missed his ongoing conversation with Athol Burns, which had begun to take on a mesmerizing suspense as it unfolded from one week to the next. At dinner Manfred didn't even get beyond the soup. "Uh-uh," Adele said, watching him gag across the table. Florence put him to bed and propped his pillow against the yellow headboard. Next to his bed, she placed a small table with a glass of water, a box of suppositories, and a bottle of cough medicine, even though he had no cough. Nor was he constipated. In Florence's medicine chest, suppositories held priority over everything, but only for twelve hours. "In the morning, if nothing's changed, we'll really clean you out," Florence said hopefully. She was holding her own belly as she stood next to the bed. "Let Uncle Lester know," Manfred answered, in a weak voice. He liked being in bed. He felt safer there than he had in months. Florence placed the back of her hand against his forehead. "Let me have a look at your eyes," she said. Adele hovered at the door, Max behind her. "I can always tell by the eyes." She didn't say what she could tell. "Maybe it was too much of a good thing," Max said over Adele's shoulder. Twice during the night, Florence got up to check on him, her thin nightgown transparent in the dim hall light behind her. But Manfred was sleeping soundly each time, wrapped in a feverish full-bodied lassitude and strange Wagnerian dreams of tall ships sailing bravely for home, for the ports of Germany.

Chapter Seven

Quarterlies came and went with the usual fierce intensity, breeding adolescent hysteria (and foul stomach ailments) in the process. There was a minor cheating scandal at Forest Park in Latin II (Caesar and the Gallic Wars), and three girls fainted in senior Chemistry in the middle of the exam, causing considerable disruption in the amphitheater-style classroom as they keeled over, one after the other. The intake of aspirin product, as well as paregoric, more than trebled among the student body, and the school's tiny dispensary, furnished with a single iron cot, was overburdened by a formidable waiting list of patients. Manfred got a 98 in History, 94 in French, 91 in Geometry, 94 in English, where his teacher allowed him to resubmit the essays written for Athol Burns for extra credit, and 82 in German. The highest mark in German was 86, Louis Lambrino being very tight, very retentive about grades, even penalizing his students for handwriting that did not please him. Manfred's handwriting, to say nothing of Jojo McAllister's, did not please him. Adele's marks almost equalled Manfred's, give or take a couple of points, and Billy Brent didn't do so badly, either. He was not quite in their league, perhaps—all those nineties—but for the moment Tidewater College was still a realistic hope. By now, fully recovered from the episode in the Kates's garage, he also had a new girlfriend, a cheerleader named Sheila Taylor, who lived out in Howard Park. Sheila could jump high, back arched nicely, at football games, and she didn't read books. That suited Billy's disillusioned mood perfectly. Billy was doing fine.

As for Jojo McAllister, he not only failed German but for good measure also managed to fail Geometry. In turn, his humiliated soul then failed him, folding up at the dreadful news like a fragile wing that has snapped in midflight. It was painful for Manfred to run into him on the way to school. While Jojo hobbled along, his body as suddenly crippled as his soul, he took to laughing sardonically out of one side of his mouth at his own stupidity, making jokes on himself that he kept repeating for effect, already convinced that buffoonery, in some or all of its variations, was his true calling. If he couldn't be a real person, he'd become a clown. Manfred tried halfheartedly to cheer him up, but it was clearly hopeless. His inane encouragement only depressed Jojo. They were becoming miserable in each other's company. One was smart, the other stupid. It was all so clear, and it was forever. The knowledge of forever meant true despair. That was when Jojo started his first ulcer.

After quarterlies, Manfred returned to work at M. Gordon & Son. It was back to normal, Max said. Manfred had graduated from delivering orders. Sometimes he was allowed to check incoming invoices, often he acted as an assistant salesclerk for small-scale items, objects like thin, gold-plated bracelets that didn't cost too much, or necklaces made from semiprecious stones, or even Schick razors, which had become one of their basic items through Max's skill at advertising. For the really big sales, for important diamonds and emeralds and rubies, Max sometimes let him observe the transaction at the counter, placing him discreetly at his side and just behind him, sober and stiff and aloof. For these special customers, who usually arrived by appointment, Max put on a custom-made alpaca jacket. Under it, he always wore a white shirt. It all went together, the beautifully stitched jacket, the starched shirt, the perfect little bow tie, as well as Max's snowy hair, honey skin, and the eyepiece stuck so professionally in his eye-socket. "Marriages are made in heaven . . ." Who would have thought? These important negotiations sometimes took an hour or more to conclude, while the customers sat on leather-covered stools on the other side of the counter, facing Max and his silent footman Manfred. Delicate queries were posed, precise answers crisply returned. There were breathless hesitations on the brink, a not-always-pleasurable suspense shared by all, then the resolution one way or the other. Money, it seemed, showed terrific resistance to change. Later, once the alpaca coat was removed in the back room, Manfred could see the heavy sweat stains under Max's arms and feel a kind of breathlessness that came from him, almost a panting, win or lose. There was a sexual fervor to it, a reek of physical effort. Max's whole body was clearly spent from exertion. He worked for his living. That much Manfred would concede.

At home, meanwhile, Florence knitted and crocheted, isolated by her new happiness. (Everyone was crocheting these days to save money.) She had turned in on herself, it seemed, with an almost perfect selfishness. She was forgetful of daily life, a little forgetful of them all, wholly satisfied to focus on the growing swell of her stomach and on the quick movements, the sturdy kicking, like a pony's hoof, really, that often came late in the evening, as they all sat in the living room digesting their supper, while Max in his easy chair read out his little newspaper headlines in the ongoing search for the world's dependable ironies.

On weekends, Bernard Potter continued to show up. B. P., as they called him, to Manfred's contempt. He was there on Friday and Saturday nights and sometimes on Sunday afternoon, too, smoking up the house with his cigarettes. "Any intentions?" Max joked to Adele, looking, however, somewhat guarded, as though he was afraid of the answer. When Bernard's slick two-seater with the rumble seat stood out front, everybody on Fairfax Road knew that it was parked there for Adele Gordon. It was nice that it was a two-seater, Adele told them, because then they didn't have to double-date with anybody else. As though somebody had raised the question. It was really nicer that way, she told them airily, combing out her hair in the living room.

On the Saturday that the *Emden* docked at the Recreation Pier, at the foot of Broadway, Manfred again found himself wandering around the waterfront alone at lunchtime, deliberately searching for the ship. It was the same old midday habit that had brought him so much pleasure until now, enhanced by a touch of sweet spring weather, a sudden Baltimore balminess that would hold for a week. The bay threw off a sharp smell of seaweed and shellfish, an astringent saline mix, that was strong enough to overcome even McCormack's spicy smells, which usually permeated the harbor area. Walking east along Pratt, taking in the pungent air, Manfred stuffed down one of Florence's indigestible egg-salad sandwiches. He was on the lookout. Black prows hovered over him. Waterlogged anchor chains hung taut from the sides of the ships. It was hard to walk on the cobblestones, they hurt the soles of his feet. A few minutes later, looking up suddenly, he finally found the *Emden*, sitting powerfully at the pier a block away, flying pennants and flags and special festive bunting. It was going to be there for ten days, the *Sun* had reported, then head for Spain, or Wilhelmshaven, someplace like that. Already, it had visited Hawaii, San Diego, Boston, and Philadelphia, sowing goodwill and friendship in every port, in the year of the Olympics, although it had chosen to skip New York on this tour, its captain had implied in an interview, because of its unpredictable mongrel population and fierce radical politics.

The *Emden* didn't look, closer up, quite as massive as Manfred remembered it from the straits at Bremerhaven, when he had sailed for America. That day, with the sun beating steadily down on the vast plain that lay behind them, all the way to Denmark, the whole crew had lined the *Emden*'s deck to wave the *Europa* on its way. At the *Europa*'s railing stood Hugo Bondi, the Heller brothers, Erwin Marks, and Manfred Vogel. Drums had sounded in strict march-time. A bugle roll curled across the water. The crew was dressed in fastidious white, standing side by side in a perfect line with their hands clasped behind their backs. A lone seaman stood high in the rigging, one leg outstretched for balance. It had all glowed in the northern winter sunlight, heightened by extraordinary emotion. Now the *Emden* seemed diminished by normal activity. Ordinary sailors ran around the decks in heavy sweaters. Officers shouted orders. Water sloshed overboard from the deck. They were getting the *Emden* ready. There would be many official ceremonies on board during its stay in Baltimore, the *Sun* reported. There was also going to be a party in the Annapolis State House hosted by the seventy-six official German societies of Baltimore, and the Germania Club was giving a supper-dance at the Vorwaerts Turnverein the following night at six P.M. And while all this, and more, was going on, the public was invited aboard at certain stated hours to roam free, more or less, taste vicarious adventure, and admire a certain spirit of Saxon efficiency that was getting short shrift in various quarters these days. At the stern now, the swastika hung limply in the spring breeze. An iron cross marked the prow. When Manfred looked closely he could see rust marks near the waterline. The *Emden* had been at sea for nearly six months. On deck canvas hoods protected the guns. Manfred heard a few German words, a few exclamations and curse-words. It was all very serious up there, like school in Frankfurt. It was hardy and unsmiling. Crucial matters were at stake, the destiny of the world, perhaps. His own father had once been part of it. So had he. It was very German.

When Manfred got back to the store, Lester and Max were standing side by side behind the counter, like a couple of querulous old partners who are a little tired of each other's company. Max was writing out a sales slip for a customer, using a pad that made instant carbon copies. Lester was watching over his shoulder, alert for mistakes. It was the way they sometimes behaved with each other, especially Lester. Now the customer picked up her purchase and left the store, murmuring thanks to Max in a pleasant voice. Another sale, an important one, in just the right environment. That's what precious gems need, Max always said. Patience. Quiet. Consideration and time. A setting of luxury to shine against, to glisten at their best. That's why their boxes were lined with black velvet, Max told Manfred.

"You're back," Max finally said, glancing up at Manfred as he tore off the sales slip for future accounting.

"Am I late?"

"No, you're not late. You have forty-five minutes like everybody else. What'd you do for lunch?"

"I had a quick bite."

"Want to come to Hagenzack's this afternoon?" Lester asked.

"It's up to the boss," Manfred said lightly, gazing at Max.

"Why not," Max said smiling. He was in a good mood from the sale he had just made. He took off his alpaca jacket, brushing off a piece of lint. The back of his white shirt was wet.

"Where'd you have lunch?" he asked, a second time.

"I was down at the docks."

"Again?"

"I wanted to see something."

"What'd you want to see?"

"I wanted to see the *Emden*."

"The Nazi ship?" Lester asked, rapping his knuckles on the counter.

"Yes."

"What'd you want to do that for?" Max asked.

"I was interested."

"I would've thought you had enough of that in Germany."

"I like ships."

"There are plenty of other ships down there," Max said in a suddenly irritable voice.

"We'll go to Hagenzack's later," Lester said, moving in quickly. "The old man says you can use some of the equipment."

"I don't have a jockstrap."

"Run across to Gutman's and buy one." Lester reached for his wallet.

The day crawled on. The afternoon's business was dull. It was that time of the year, long after Christmas, not quite Easter. Manfred and Lester walked over to Hagenzack's at three o'clock. Upstairs, the old man let Manfred fool around with some antiquated weights that no one else bothered with. He also allowed him to use the steam-room, a major concession, first speaking a few condescending words to Manfred in German to keep from getting rusty. Then he moved off to encourage the real customers. In the steam-room, there was talk of the *Emden*. The visit was not welcome, and voices were bitter. There was a suggestion to picket. A general boycott was thought desirable. There was another suggestion to picket City Hall, too, because of the mayor's official welcome. Someone then said that old man Hagenzack had invited the *Emden*'s captain and his second-in-command up to the club for a free hour of exercise, in case life on dryland began to make them feel sluggish. "He wouldn't dare," someone else said in a threatening voice, and everyone who was Jewish agreed. The steam poured in, thinning as it rose in the air, the naked bodies, wrapped in sheets, lay back on stone slabs and sweated. There was barely room for Manfred. Judge Sandler, from the Orphan's Court, occupied a whole slab to himself, his great stomach rising high above the rest of him, soaring like a great red mountain. It was Judge Sandler who had said, "He wouldn't dare." Sally Krieger's father reclined on the other side of the room, his belly button disappearing into his stomach like a tunnel. He greeted Manfred with a pained nod; Sally Krieger's father hated sweating. There was another mumbled remark about the *Emden*, further angry objections, all instantly diminished by the heat. Off on his own, Lester looked miserable; his eyelids began to droop, he kept snorting into the steam. Sweating in the corner, Manfred made sure to keep his mouth shut. He contributed nothing about the *Emden*. He didn't explain about Julius Vogel and the German Navy. Nobody was interested in that kind of information, anyway. It was too personal.

It finally ended when Lester and Manfred headed for the showers together, towelling themselves down vigorously afterwards and slapping on talc and Aqua-Velva from a communal shelf. The wiry hair on Lester's shoulders bunched together like iron filings. "What on earth got into you?" Lester asked as they were dressing. "What do you mean?" Manfred said. He hid himself behind the coarse towel, reaching for his underwear. "Visiting the *Emden*," Lester said. "I just felt like it." "Turn around and look at me. I can't talk to your back." Manfred faced him. "They're only here for propaganda," Lester went on. "You should know that better than anybody." "I do know that." "So?"

"We'd better hurry," Manfred said, slipping into his pants. "Supper's early on Saturday and Adele's going formal with B. P. tonight." It was a lie. Adele was going formal next week.

"Oh, Adele," Lester said, tieing his shoes. His balding head was shining. His cuticles were perfect. He began to hurry a little, slipping on his tie, making a Windsor knot that he eyed for imperfections in the mirror. The *Emden* was forgotten, at least by Lester. It was almost five-thirty, time to head uptown and dress for the Ottingers' weekly dinner at the Suburban Club, where Lester could pretend to snub his old cronies.

A few days later, Manfred left the store without telling Max and headed down to Pratt Street again. It was late afternoon, after school. There was the same scent of seaweed and shellfish coming from the bay, the same mixture of ship's refuse and fuel. At Pratt Street, where the traffic was heavy, Manfred turned left. In the harbor, rainbow slicks floated sluggishly by. Rotted lumber gathered at the water's edge, old inner tubes streaked with seaweed rose and fell in the wash. This time, he made it to the Recreation Pier in less than five minutes, scurrying as if he was being chased. At the pier, he headed for the sentry on duty, without hesitating, and, after picking up a pass, took his place in line to board the cruiser, along with dozens of other visitors. The flags still flew in the spring breeze, the pennants flapped overhead. It was open house, extending from three o'clock to five every afternoon. Below him, as he slowly climbed the gangplank, waiting his turn, picketers marched in an unsteady line, three middle-aged gentlemen from the Maryland League Against War and Fascism and a single emaciated student from the Johns Hopkins University. They all carried crude signs that read "We Protest!" One of the middle-aged gentlemen was bellowing at the visitors boarding the ship. "Nazi-lovers!" the man shouted. At the sound of his voice, the sentry below stiffened at his post and everyone began to talk louder. "Hitler, the Enemy of Mankind!" the man yelled. Alongside him, a cop stood self-consciously on duty, hefting his nightstick in the palm of his hand.

On board now, crew and visitors were eating *streusel kuchen* from a long buffet set out on the main deck. Two cooks, wearing long white aprons, did the serving, urging snacks on guests with unctuous attention. The smell of raisins was in the air, citrus fruit, sweet Moselle wine. A tall vase filled with gladioli sat in the middle of the buffet. Manfred didn't wait for refreshments. He wanted to separate himself from the crowd. He paused a moment to get his bearings, prow to stern, then hunched over as though to make himself less visible, he looked left and right, looked again, and headed below, slipping unseen through an open hatch. He had never been on a cruiser before. He had never been on any naval vessel. All he knew of life at sea was third-class on the *Europa*. The *Emden* was not like the *Europa*. It amazed him, as he proceeded along an empty passageway below, it was so clever. His father had always told him that German naval ships were masterpieces of cleverness. His father was always bragging about it. They were more clever than British ships, he claimed, because they couldn't take anything for granted. On the *Emden*, nothing was what it seemed, or rather everything was what it seemed and something more. It was full of exquisite space-saving devices and other things, a masterpiece of German ingenuity. Nothing aboard, Manfred saw, was loose. The crew slept four high in bolted bunks. Immovable lockers stood in stacks. Everything was moored, held tight by a floor, a ceiling, or a wall. You ate standing up if you were a member of the crew. You stood at a high table-counter, which was bolted to the floor. That way you eliminated the need for chairs. Then he saw that the toilets were lined up face-to-face in the prow, starting right at the sharp angle—the knifelike jut—at which the two plated sides of the *Emden* were welded together. There were no partitions between them, making primitive camaraderie an absolute necessity; no one could sit on a toilet without talking to the man across the way or alongside him. That in itself was very clever.

Manfred rushed on, through a smell of oil and metal. He heard voices above him, steely echoes, strange bells, stiffening himself against the wall at each sound. He prodded a storage door open, saw buckets and mops. He seemed to know where he was going, without ever asking the question. He was used to rushing like this, it was one of his oldest habits. Finally, in a gangway near the stern, a sailor grabbed him by the arm and said something. "Ver-bo-ten," he said through his teeth, enunciating carefully. He held tight to Manfred's arm, just above the elbow, his thumb pressing hard into the flesh. Manfred smiled complacently at the sailor before answering. As he explained himself, at length, the sailor let go of Manfred's arm. "Where did you learn such beautiful German?" The sailor had a Swabian look, with a turned-up nose and blue eyes, like Joachim Adler, the butcher's son. Manfred was off-limits, the sailor said politely. It was regrettable. This deck was closed to the public. There were signs everywhere, yes? The sailor pointed to starboard and port. They could not open the whole ship to just anybody, he added. Manfred and the sailor had a brief exchange then. It was very friendly, man to man. The sailor was only eighteen, he told Manfred, the youngest of three seamen in his family. He came from Bavaria, not Swabia. He had not seen Germany in half-a-year, he said.

An officer squeezed by in the narrow gangway, brushing against Manfred. His uniform was covered with gold buttons. He said something to the sailor, asked a question, then joined in the exchange. But yes, he confirmed politely, Manfred was off-limits. How could he be otherwise? After all, the *Emden* was German. It belonged to Germany, not to the public. Would the American Navy, he asked, give ordinary German citizens the right to explore an American vessel as they pleased? Of course not, the officer said, answering his own question. The officer looked like Kurt Vogel, with the same luminous inquisitive eyes and tense, thin mouth. He held his back rigidly straight as he stood there on his own ship, like Kurt in the most ordinary situations. His shoulders were up, his spine stiff. That was how his generation—and Kurt's—had been taught in school. He was clearly very intelligent. His German was absolutely pure, without the slightest taint of dialect. Manfred began to show off for him then, like a baby brother. He tried to think of a joke to tell, tried to act bright, beginning to prance the gangway precociously, a dubious characteristic that had always set some people's teeth on edge. Even Kurt Vogel couldn't stand it when Manfred began to act precocious. The officer held up his hands. Enough, he suggested. But he was more than kind about it. He said wait a moment and disappeared up a hatch. The sailor who looked like Joachim Adler folded his arms and leaned against the wall, without speaking. The gangway was so narrow that Manfred thought he could feel the sailor's breath in his ear. It was like singing alongside Billy Brent in Miss Tubman's music room. When the officer returned a moment later, he was carrying a glass of champagne. He offered the glass to Manfred.

"Oh, no," Manfred said. "I have to go."

"Ridiculous," the officer said, while the sailor from Bavaria nodded agreement. "You are a guest."

"Well, maybe just a little," Manfred said, taking the glass and sipping the champagne slowly. It was tepid and thick-tasting, the color of urine. He checked his watch then, wondering what they were doing in his absence at M. Gordon & Son.

"There are many people who speak German in Baltimore," the officer said. His eyes were as intense as Kurt's at his most passionate.

"Perhaps," Manfred said, taking another sip.

"Yes," the officer said. "Even the protest line. You say protest line?"

"Picket line," Manfred said. A clear tide of warmth was rising from his chest through his throat into his cheeks. He could feel it in his temples, too. He sipped again.

The officer smiled conspiratorially. "You have seen the protest line? It is nothing, of course." He winked at Manfred. "Some Jews," he said.

"I suppose," Manfred answered. As he spoke, he frowned without knowing it, his forehead, just over his eyes, coming together unpleasantly in a knot. There seemed to be a spiderweb strung stickily in the warm air above him, coming between him and the officer and the sailor from Bavaria. It was like a barrier that had to be eliminated. He reached up to brush it away. There had been a spiderweb in Mrs. Kates's garage, too. Hep, he heard somewhere behind him. It was like being in the Schillerplatz. Hep, he heard again. He tried to think of a joke, but they were already laughing at something he had just said. What he had said was, "Your name's not Joachim Adler, by some chance? You're a spitting image." The Bavarian boy was laughing louder than the officer. Manfred began to laugh with them, it seemed so strange at the moment. As he laughed, two red marks appeared on his cheeks. The air was very bad in the gangway, the smell of oil was everywhere. Manfred listened to the vague hum that came from somewhere in the stern, a steady, ongoing rumble of engines and turbines. That was where the oil smell came from. That was why the air was so foul in the gangway. That was not so clever of the Germans. But he had to go. There was no choice, they'd have the cops out at M. Gordon & Son if he didn't get back soon. They'd be patrolling the streets, searching for him. "Young exile disappears in midafternoon," the headline in the *Sun* would read. Florence would grow hysterical, Adele would be sorry she ever. . . . "Nein," his Bavarian friend protested. "Stay." Manfred finished the champagne, ate a cookie held out to him by the officer. The officer bowed from the waist. His cheeks were like a girl's, not a bristle showing, tawny and smooth. Manfred was smiling. He was happy there, enclosed by metal and oil and the two German seamen. He told himself that he was happy, saying it in German. That seemed perfectly natural. He wanted to ask them what was happening in Germany, what was really going on in Frankfurt, in the street off the Mendelssohnstrasse, but he kept the question to himself. When his new friends insisted, he promised to come back before they sailed. They were all still laughing together, crowded there in the gangway in the fetid air. He promised to return soon.

Over the spaghetti and tomato sauce that night, Florence told them all that she had discovered a stain that afternoon.

"I could hardly even be sure," Florence said. "It was just a drop, barely that. A red stain. I don't want to make it sound too important."

"We're going on a new regime in this house," Max announced.

"Mummy," Adele said. She had been calling Florence "mummy" ever since she had discovered that she was pregnant. "Mummy, we'll get somebody to come in every afternoon to clean. Like Laura's mother does, like Mrs. Piscitelli."

Florence settled back in her chair and pushed her plate away. She sat on the base of her spine, belly thrust up; on her feet, which were sometimes swollen now, she wore an old pair of felt slippers that had belonged to Max. It was Manfred's opinion that any red stain was probably caused by the debilitating quality of her own cooking. There was so little nourishment in Florence's supper menus that bleeding resulted. It was the body's protest against deprivation. He was probably bleeding somewhere himself.

"And we'll all start making our own beds," Max said. "You hear that, Manfred, every morning you're going to make your own bed like the rest of us."

He nodded and sucked up a few cold noodles. At the side of his plate there was a little lake of red sauce, where he had collected lumps of tomato skin and the watery liquid in a corner of their own.

"Did you call Dr. Sacks?" Max then remembered to ask.

"Of course. He always told me it was a possibility. I always knew that. It's the dangerous time." The presence of risk seemed to make her glow.

"That's why we have to make some real rules around here."

"Mummy, I'll do supper from now on."

"Oh, Adele, don't be ridiculous."

"It's going to be a new regime around here," Max said, making a fist.

"Dr. Sacks says it's just the uterus adjusting to the weight of the baby. It's just a sign of a little strain, is all. It happens all the time. Sarah Kates was in bed with Hilary for four months when she was carrying, remember?"

"He wants you to go to bed?"

"Mummy, you go to bed, I can do breakfast, too."

"Well, I'm sure that's all very nice," Florence said. "But it's not all that serious, just a couple of weeks of touch and go."

"Touch and go?" Max asked. "Is that what Sacks said?"

"It's that time. It happens to everybody."

"I'm going to fix the coffee," Adele said. "Who wants coffee?"

"I'll help you," Manfred said.

"Just stay put. I can handle a pot of coffee myself."

"You always want me to stay put. I can't make a move around you."

"Come on, kids," Max said in an aggrieved voice.

While Adele fussed in the kitchen, an air of heavy piety settled over the dining room. Florence carried her self-righteous look, Max appeared grave. It was the old one-two. What was a drop of blood, anyway? Touch and go, Florence had said, batting her eyes at the table with pleasure.

Manfred began to bang his knee against the leg of the table, rattling it nervously. A glass shook in front of him.

"Stop that racket," Max said.

Poor fool, Manfred thought, holding on to his knee to quiet it. That was all Max was. Poor ordinary piece of human banality. A fool. And he didn't even know it.

"And finish your spaghetti, food costs money."

It always came down to that.

"If you're looking for something to do," Adele called from the kitchen, "you could start clearing the table."

"Maybe you could change all the sheets tomorrow," Florence said to Manfred after some thought. "Tomorrow is fresh linen day, you know."

Manfred glanced up at Florence. She looked happy. She had what she wanted. One drop of blood and she was crowned like a queen. Mummy, he thought. Shirley Temple talked like that. It was her favorite word. Mummy, with a perfectly calculated pout.

The subject, Athol Burns said, was the emanatory elements of life. At least, that was what Manfred thought Athol Burns had said, but he couldn't be sure. With Athol Burns you had to listen with your whole body and something else, as well; you always had to be on full alert. Strangely, it was a condition that Manfred felt totally comfortable with; he was at full alert all the time now. Of course, Athol Burns went on, you have never heard of the emanatory elements of life. (Speaking in dismissive but almost proprietary tones, hands folded properly in front of him.) Athol Burns was right. Manfred had never come across the subject in his readings. Nor had he ever heard anyone speak of it. He blinked now as he tried to spell the word to himself. Did it begin with an *a*? An *e*? Perhaps an *i*? In Baltimore, *e*'s and *i*'s sometimes sounded like the same letter. Imminatory.

Probation was still in effect. It was entirely natural by now. Athol Burns could not give Manfred up. They still faced each other across Burns's desk for forty-five minutes each week. It was an established liaison by now, full of interest and concern for the assistant principal of boys. Athol Burns allowed no interruptions during these sessions. They were his property alone, staked out like promising lodes of rich new ore. Both Miss Ficker and Mr. Lund knew better than to break in during their exchanges, no matter how serious the reports of disturbance in Boys' Study Hall might be. As for Manfred, he was now perfectly at home with Athol Burns. He knew what to expect, as he knew what to expect from all his teachers. He went from one teacher to the next almost without thought, deftly changing coloration along the way like a lizard darting against the background of sometimes threatening flora. For Athol Burns, he tamped himself down, his voice went flat, he forced himself to become a neutral grey and white; and he had learned to listen with absorption. They were a match for each other.

"There is not another student in this school with whom I could discuss such a concept," Athol Burns said over the pyramid his fingers made. He was referring to the emanatory elements of life. "The level of maturity forbids it. We have few real young adults among our student body. But I must not complain. It probably could not be otherwise. Beggars cannot be choosers, and so on. I trust you follow my line of thought. I trust you recognize my meaning."

"Your meaning," Manfred said, in the incantatory style they had both grown used to in the last few months.

"Of course, when one raises the question of emanatory elements, one must immediately add that its opposite, that which I choose to call the sterile sources, is equally powerful. Only a fool would deny that. It is the adversary of adversaries. You in particular understand that, Mr. Vogel. Tribulation strikes deep and leaves black wounds which do not always heal properly. And tribulation is always there, lying in wait for all men. But our real strength thrives on the giving side, where sweet generosity is the source of all good among men and soothing beneficence the only true gift in life. I choose to call this giving side the emanatory element. On its behalf, the juices must be made to flow. You will recognize that it is both spiritual and physical in essence. It incorporates psyche and body. It encompasses soul and soma. It is simply what emanates from every one of us at our best, in all forms, to nourish the object, to fructify, to expand."

"Fructify," Manfred said as Athol Burns waited. The subject seemed a little dimmer, more obscure, than usual.

"You get my meaning?"

"Fertility?"

"Fertility and something more. Fertility and much more. Fertility is first of all the quality of being receptive. In the broadest sense, of course. It needs the implant, the seed. In my lexicon, it is entirely passive. It is always waiting to be seized. By itself, it is powerless. No, I speak of fertility and something else. For want of a more precise expression, I call it emanatory elements. What emanates positively from man. And woman, of course."

"Yes."

"Eventually you will emanate from Forest Park High School."

The statement took Manfred by surprise. It had not occurred to him in the context of Athol Burns's words. Something in the exchange seemed to have become twisted, to have suddenly taken on a whole new meaning. He was not prepared for it. He sighed profoundly, gazed abstractedly out of the window behind Athol Burns. "Yes, sir," he said.

"You are not focusing," Athol Burns said.

"I am thinking about emanating, sir."

Athol Burns looked at him a little sharply. "And what are you thinking?"

"I'm not sure yet. I have to consider the idea. I never thought of it before."

"The point I am coming to is that I want you for Harvard." Athol Burns unfolded his hands and placed them on the arms of his chair. His coppery face became darker, as if a fire smoldered beneath the skin. "Forest Park needs a Harvard connection. We have never had a Harvard connection."

Manfred sighed again, looked blank.

"You get my meaning?"

"Harvard, sir."

"We have slightly more than two years. It will be done."

"What?"

"We will prepare you for Harvard."

"College, sir?"

"Mr. Vogel, we come back again to the question of your attention span. It is a weakness, as I have stated before. Quite serious, quite consistent. You do not seem to have the ability to hold onto the subject. You do not exercise enough concentration. Where is your will power? You are not even listening to me now. I do believe that with all my heart. Stop that awful grinning. This is a perfectly serious matter. Do you think it's every year that we get a student like you? There must be a planned course of action. We will trace it together. We will be collaborators in a grand beneficent conspiracy. We have the time."

"I'm not sure, sir."

"You're not sure? But I'm not asking for assurance. Obedience is what I want."

"Obedience."

"A simple idea. Think about it."

"I thought College Park."

"Think again."

"Sir?"

"Raise your sights, son."

"It's so cheap there."

"Cheap? Cheap is not the issue. The University of Maryland is for the run-of-the-mills. This school and all other schools are filled with run-of-the-mills. You are not a run-of-the-mill. Run-of-the-mills are a dime a dozen. That you must learn to believe. Let us circle back to our original subject. It is the starting point for our venture. The original subject is the emanatory elements of mankind."

"Emanatory elements, sir."

"Acts of creation, acts of procreation."

Manfred raised his eyebrows.

"Mr. Vogel, free your imagination. Do not limit your understanding to received notions. I go beyond fucking. I face the absolute frontiers of giving."

They stared at each other possessively. Athol Burns's axe-face challenged Manfred. There was nothing Manfred could do about Athol Burns. Athol Burns was an expert on emanatory elements. He was an emanatory element in himself. His attention span was immense. It was without end. Athol Burns knew how to hold onto a subject. He would never let Manfred Vogel go. He would have his way. Manfred Vogel had become a chosen one, in the grip of a divine force: a Burns prodigy. His juices would surely flow on the way to Harvard.

Chapter Eight

On Saturday morning, Manfred folded an extra handkerchief in his pocket, one of Benno's dozen blue-threaded gifts, initialed LV. Then he took out Dr. Schwabacher's blue vial and stuck it in a pocket of his windbreaker. He was also wearing his expensive knickers, from the Schillerplatz, which he had kept lying in a bottom drawer since his arrival in Baltimore. *Alles Vogel*, he told himself; *alles* Deutsch. He could hear Adele stirring next door, heard her cough and mutter something incomprehensible to herself. It would take her an hour to come fully awake. His bed was made. Breakfast was ready downstairs. Max was setting the table. On Saturday, while Florence and Adele slept late, enjoying the benefits of the new regime, Max prepared hot cereal, great thick lumps of oatmeal sunk in boiled milk like pieces of coal. Manfred checked his wallet, patted the handkerchiefs, sniffed once at the blue vial. He had eighteen dollars in the wallet, his address book in his windbreaker along with a few tattered letters from his mother and father and Kurt. His knickers had become a little tight in the crotch, in only six months. At the last minute, while the opium scent floated headily above him, he stuffed another one of Benno's handkerchiefs in his pocket. *Alles Vogel*, he told himself again; *alles*.

Downstairs, they mumbled at each other over the cereal, trying to swallow the lumps without choking. Max rattled the newspaper in front of him. There were pouches under his eyes, puffed olive bags that stood out from the rest of his face like tender little balloons. He looked like he was dying for more sleep. Time to go, he said, sipping at his coffee. He ran his fingers through his hair. Upstairs, one of his hairnets sat on his dresser. He had two, so he could change off from time to time. They had a tendency to snag, and Max liked his hairnets perfect. He tossed the car keys at Manfred. Why don't you get it warmed up, he said. I'll be right out. It was early April, but it was still cold in Baltimore. Everybody complained that it would be a late spring.

They would all be cheated, Max said, they would go from cold to hot overnight and everybody would get pneumonia.

Sitting behind the wheel a few minutes later, Manfred held the gearshift in his right hand. The marble knob was vibrating inside his palm. While the exhaust pounded away, Mr. Piscitelli suddenly drove past, hunched over the wheel of his car. His head was thrust forward nearsightedly, his nostrils dilated with tension. Already he was going forty miles an hour. He had probably chained Laura to her desk, Manfred decided, stuck her away with only a crucifix for company in an upstairs room in the Piscitelli house. Catholics. Priests. Nuns. And Jesus Christ. It was hard to be a Catholic and wear all that Faith, like clothes you were never allowed to change.

Manfred still couldn't get it straight. There was a Father, a Son, a Ghost. And other strange things, including church smells. How could anything so simple grow so complicated? Who was Who? And Where and How? As far as Manfred could see, all anybody wanted was to live forever. That was all he wanted. He checked his watch now. Across the street, Mr. McAllister was getting into his little jalopy, a black model-T with a rumble seat. Watching him, Manfred raced the motor a little. The noise swelled over the model-T's, carrying with it a humming that traveled from the accelerator through the sole of his foot into his shin. He raced the motor again, it felt so good. The marble knob shook in his hand. Already, he knew how to drive the car, just from watching Max. Across the street, Mr. McAllister chugged away to his nonexistent job. Nobody on Fairfax Road knew where he went every day, not even Jojo, not even Jojo's mother. A stranger came down the path then from the Gordon house. He was wearing a wool cap and a muffler that hid his lower face. His hands were gloved. His collar was up. It was Max. As he pulled open the door, waiting for Manfred to move over, he said, "The paper calls for snow. Can you beat it? In April?"

They stalled only once this morning. It was probably because it was Saturday and traffic was light. When traffic was light, Max seemed to get the hang of driving. Around Lexington Street, Max took a couple of minutes to find a parking place that pleased him, then puffing steam into the air together, Manfred and Max made their way to the store. "Here," Max said, handing over the keys, "you open up and I'll pick up some coffee. You want some coffee?"

It was cold inside the store. Waiting for the heat to come up, Manfred turned on the lights and reached for a duster. He swiped at the counters and a few open shelves. Max would check out the safe and all the safety devices when he got back with the coffee. Across the street, Mr. Levinson the optician was turning on the lights in his store, and Manfred could see the salespeople from O'Neil's hurrying up Lexington Street to their jobs. A trolley screeched turning the corner, someone Manfred didn't recognize waved through the window of the store. The day would start slowly. It was Saturday, when there was always an extra hour for everybody.

By noon, three customers had come into M. Gordon & Son. One bought a watch, another a gold bangle, and the third was just looking around. At eleven-thirty Florence called to remind Max to remind Manfred to pick up some undershirts at the May Company. There was a sale on the third floor. "Never saw those before," Max said, eyeing Manfred's knickers as he hung up the phone. "I brought them from Frankfurt," Manfred said. After a while Max ran around the corner to visit one of the radio stations. "Marriages are made in Heaven . . ." was to be rescheduled on the air still another time, and the arrangements had to be reviewed again. That was one of the things about M. Gordon & Son that Max enjoyed most. Another customer dropped in while Manfred sat on the high stool behind the counter, waiting for business. This time he sold a Schick razor, carefully filling in the sales slip and wrapping an especially neat package with white ribbon. Then he sat on the stool again. After waiting another few minutes, Manfred reached over to his windbreaker, took out Dr. Schwabacher's blue vial, and sniffed at it for a second time that day. Then he tasted a drop. A spurt of something sweet instantly went through him. He began to hum at his place. He began to rehearse his part in the Boys' Chorus, humming high above the melody that Fatass Crimmins would croon. His falsetto sounded good to him in the empty store, circling somewhere above him near the ceiling. When Max returned from the radio station with a lunchbag in his hand, Manfred was singing aloud, but just barely. "I didn't bring you anything," Max said, "because I didn't know what you wanted. Why don't you hop up to Huyler's for lunch. I'll treat."

That suited Manfred. "OK," he said, taking a slow step on the carpet, then another that seemed to go on for five minutes. It was the opium, he knew, repositioning the world, just as he wanted it. He put on his windbreaker. "I sold a Schick razor," he said. "To a lady."

"Good."

"She paid by check."

"Did you ask for identification?"

"She showed me her driver's license."

"Always ask for two identifications. Always make sure there's confirmation."

"She looked honest."

"That's how you go broke, depending on looks. Here, take this." He thrust two bills into Manfred's hand. "Take your time."

Manfred walked over to the May Company and bought two undershirts for a dollar twenty-nine, charging them to Max Gordon. Then he walked back along Lexington Street, checking out the movies. He could distinctly hear a humming sound following him. It followed him and was pleasantly inside him at the same time. *Little Lord Fauntleroy* was playing at the New. Freddie Bartholomew looked prissy in the stills outside the theater. Freddie Bartholomew always looked prissy. He had long hair and curls. Down the street, at Keith's, Boris Karloff and Ricardo Cortez were starring in *The Walking Dead*. Some stars, Manfred thought, passing the theater by. At the Century, near M. Gordon & Son, was *Rose Marie*, with Jeanette MacDonald and Nelson Eddy, who, according to "The Talk of Hollywood" in the *Evening Sun*, might or might not be going blind. "Oh, Rose Marie, I love you . . ." Manfred sang, before turning into Huyler's. He really liked that song. At the fountain, he ordered a chocolate soda with whipped cream and when he had finished it, picked up a Hershey bar and some gum drops at the cashier's desk. That just about took care of one of the bills.

"Enjoy lunch?" Max asked a few minutes later.

"Grilled cheese and bacon," Manfred said.

Max nodded. The pouches under his eyes looked like sweet hazelnuts. He sat on the stool, gazing sleepily at his merchandise. The room was soft and inviting. Manfred sank into the rug. "Not much today," he said.

"Sometimes it hardly pays to open up," Max said.

"What do you think happened to Uncle Lester?"

"Maybe he went to the club for lunch."

Max lighted a cigarette, looking guilty. Manfred picked up the duster and went to work on the open shelves again. He was moving very slowly, trying to look as normal as possible. "When you finish that," Max said, "maybe you'll take the afternoon off. You could take in a movie."

"Maybe I will."

"Or you could go to the library."

"I'm still reading *Lost Horizon*."

"Well, don't hang around just for the sake of it. You can take the 32 home if you want. You know how to get home on the 32, don't you?"

"I could take the 31."

"The 32's faster."

"If I go to the Hipp, I'll take the 31."

A few minutes later, standing in the unheated lavatory in the back of the store, Manfred went to work on his part. He combed his hair straight forward, wetting it a little with spit, then carefully made an arrow-straight line down the other side of his head, the way it used to be. Quickly then, he combed both sides flat, tilting his head to one side to get a good look at himself in the mirror. It was the old Manfred he saw, the way he had looked before he tried to acquire the Gordon aura. Thin oval face, flushed now with self-absorption, ruddy cheeks, in the European style, blue-green eyes glinting with anticipation like certain parts of the sea at dawn. He was satisfied with that look, it made him smile at himself provocatively, for the moment it gave him back the Vogels, whose faces he could now clearly make out in his own. When he came into the front of the store, slipping into his windbreaker, Max said, "Here's another dollar. Add it to the change from Huyler's. It should cover you for the afternoon. What were you doing in there?"

"Nothing."

"I didn't hear the toilet flush."

"I was combing my hair."

Max examined him from behind the counter, lighting another cigarette. Making an O with his mouth, he began to blow smoke rings, a trick he had recently learned from B. P. Potter. "Well, have a good time. Don't be late for supper. Adele's going formal tonight."

"I hope it doesn't snow. B. P. doesn't like to drive his car in the snow."

"The sun is peeking through."

"Should I stop back later in case it gets busy?"

"It won't get busy. It's one of those days."

"Please bring my undershirts home."

"OK. Don't get lost now."

They said good-bye then.

Once outside on Lexington Street, Manfred wandered slowly up the hill towards the Century. The humming noise still followed him. Nelson Eddy was wearing a Northwest Mounted Police uniform in the stills out front. He was holding Jeanette MacDonald in his arms, and they were both staring straight at the camera with their mouths open. Inside the Century, everybody could watch Marvin Manning's organ rise between shows. That was another good American joke, although not quite as good as George Washington, the father of his country. Oh, Rose Marie . . . Mouthing the lyrics to himself, Manfred moved on, his hands thrust into the pockets of his windbreaker. Next door he could smell Huyler's chocolate drifting in a great sweet mass through the revolving door (following his lunch, this left him a little nauseous); then, at the corner of Charles Street, suddenly decisive, he headed downtown, past O'Neil's. Walking along like that, still smelling the chocolate, still tasting it, in fact, he took inventory again: three handkerchiefs, a blue vial, twenty dollars and fifteen cents, an address book, a half-dozen uninformative letters. It seemed more than enough. He patted his wallet, ran his fingers around the edge of the address book, looked up at the sky. As Max had said, the sun was peeking through. There was a haze overhead, dimly filtering cold yellow light everywhere. It was better than snow.

At Pratt Street, there was the familiar sharp harbor smell, the ancient acrid odor that floated over lower Baltimore as though it was being perpetually stirred by an invisible hand. It never cleared, never thinned, it wiped out Huyler's, it wiped out everything. Glancing down at the bay water, he sniffed once or twice; the odor pinched his nostrils, his nose twitched in disgust. Then, making a face, he turned east again, walking flat-footed on the slippery cobblestones. Dirty foam gathered at the water's edge below, green sludge from the Sparrows Point steel mills, human turds and used condoms, rotted garbage, rainbow streaks of oil, the sea's dregs, the world's stench. (There was nothing like this in Frankfurt, he told himself with an expanding sense of superiority, nothing so filthy, so haphazard, so humanly offensive; it would not be allowed.) Manfred sniffed again, turning his face away from the wind. Above him, the iron-black prows soared powerfully, the great anchor chains stretched to their limits. Mounds of cargo stood alongside the ships. Across the street, ancient redbrick warehouses and depots faced the water. There was a rush of trucks along the roadway, even though it was Saturday. Manfred was walking faster now, consciously pushing himself. It suddenly seemed so far to go, as though he had never made the walk before. It seemed an impossible distance. A couple of minutes passed: half a lifetime, he thought. Finally, out of breath, nose running, he reached the Recreation Pier. The *Emden* still stood at the dock, full complement aboard. Crew: 653 men. Displacement: 5400 tons. Draft: 17 feet. He could relax. The cruiser had not sailed for home. He would try to quiet down. He had promised to come back to see his friends, whose names he didn't know, whose faces he hardly remembered, except for "Joachim Adler," and even his simple Swabian peasant features had taken on a floating dreamlike quality that was as elusive as everything else. It was the *Emden*'s last day in port, *auf wiedersehn* to Baltimore, Maryland, then on to Wilhelmshaven, or another one of those shingled coves somewhere on the North Sea, where the German Navy, which took nothing for granted, had its home base facing England.

For a second time that week, Manfred took his place in line before climbing the gangplank. Down below, the picketers had disappeared. No one marched today, no one protested. Still nervously swinging his club, the cop on duty rested lightly against a hawser, looking anxious. On the gangplank, family clusters gathered in front of Manfred and behind him, stout mothers and fathers, small impatient children bundled up against the cold. Some of them were chatting in German. On the stern of the *Emden*, Manfred could make out the swastika. The sun was breaking through now, he could feel the air slowly warming. It was April, after all; spring had to come sometime. On deck, it was more crowded than he had expected. There was a throng, a crush of complaisant visitors, easy to get lost in. The *Emden*'s captain stood at the head of a receiving line, making hearty jokes and simpering in a pleased way when his guests laughed at them. But they were all laughing, visitors and crew alike. It was easy to laugh today. As the *Sun* had reported, Baltimore had been one of the most successful visits the *Emden* had made on its long cruise around the world, and Captain Bachmann was very happy. His mission was fulfilled. There had been many parties, many exchanges of goodwill. In Baltimore alone, there were seventy-six official German societies, all of them welcoming, as expected. To show his respect, Mayor Jackson had spent a half-hour on board. The German military attaché had come over from Washington to be part of it. The whole crew had toured Baltimore by bus and finished the day by swilling beer at the Deutsche Verein downtown. Today, Captain Bachmann was wearing a dramatic blue-and-white uniform with decorations strung in three meticulous lines across his chest and gold-and-silver braid on his shoulders. "*Aber ja*," he was shouting at that moment to his gunnery officer, who was on special duty during visiting hours. His officers, starting with Commander Tepp, were bending from the waist to kiss the hands of Baltimore ladies as they passed through the receiving line. All the officers looked alike, dressed in blue and gold, wearing glossy black shoes that reflected the light like mirrors. Straightening up, they directed the Baltimore guests, the prosperous mothers and fathers and fat blond children, to the champagne and Rhine wine and *streusel kuchen*, perfectly set out for the *Emden*'s last open house to show the world how these things should be done.

Manfred helped himself to a piece of kuchen, keeping an eye out for his friends. He was looking for a pug nose, a Swabian face from Bavaria. Someone handed him a glass of wine. "*Danke*," he said, sipping at it tentatively. He was being shoved around at the buffet table. Everybody was after the champagne and the *streusel kuchen*. After the third sip, resting against a ventilator, Manfred tipped his glass back and drank the wine down in one gulp. It had a nice spicy flavor, and the emerging sun helped. He heard German being spoken behind him, then English. Someone was even speaking Polish. Captain Bachmann was laughing uproariously, the gunnery officer stood behind him, chin up, awaiting orders. The ship glistened in the sun, battle-grey and shining. To the northwest stood Baltimore's skyscrapers. Their silhouettes, mostly flat-topped, took on a hard edge now in the new sun. Manfred stood there, glass in hand, shading his eyes, as he gazed out at the foreign city beyond.

With his second glass of wine, Manfred helped himself to more kuchen. It was crisp doughy stuff, better than his mother's. There were more raisins, for one thing, and there was powdered sugar sprinkled on the top. His mother always went easy on the powdered sugar at home for fear of diabetes. Backing away from the table, Manfred stepped on someone's toes and excused himself in German. Over near the receiving line, a child was crying. Crew members whispered in each other's ears, eager to set sail. There was more laughter from Captain Bachmann. The German military attaché from Washington was now standing next to Captain Bachmann. He was wearing spurs and all his decorations from the Great War. It was all absurdly polite. No one tried to talk to Manfred, no one paid the slightest attention. That was how he wanted it, it suited him. If the sun stayed like this, getting warmer every minute, he would have to take off his windbreaker, he would be able to walk around the *Emden* in his new sweater, showing off the Indian design; this one had grey-and-white diamonds all over it. He still hadn't found his pug-nosed friend, and the officers all seemed hopelessly alike in their blue-and-gold uniforms. They stood in line, kissing the hands of American ladies, looking satisfied with themselves.

While Manfred was slipping some cookies into his pocket, the seamen on deck suddenly came to attention and then stood with their legs set a foot apart, their hands clasped lightly behind their backs. Everything instantly became serious. A somber look came over Captain Bachmann's heavy face and settled in; he faced the crowd intently, composing himself for a bout of sentimental metaphysics. The gunnery officer behind him saluted. The officers clicked their immaculate heels and the German military attaché carefully brought his spurs together. An order was shouted. The crowd became silent, shushing the children. They began to move in to get a better look. The seamen were lined up as they had been when the *Europa* sailed past, coming out of Bremerhaven. They were in perfect order, as always. Their officers moved in front of them, taking up their own positions. It was as though the end of the world was going to be announced. As Captain Bachmann opened his mouth to speak, Manfred slipped through a hatch, just as he had done a few days before, and edged his way below deck. A soft spiral of dizziness ran through him as he climbed down the ladder. It was the blue vial. He could hear a speech beginning above his head, out in the open, a slow rolling acknowledgment of the glories of Baltimore and its many loyal Germans. The ship grew even more silent as the speech rolled on. *Kultur* was on Captain Bachmann's mind today, *Kultur* and *Volk*, not politics. He had made the speech many times before, with certain variations to account for time and place, and was very sure of its effect. By now, Manfred was snaking his way along the skinny corridors below. The dizziness held him. The oil smell returned along with other smells; it was exactly the way it had been a few days before. Quietly then, as the voice on deck grew inaudible, Manfred squeezed himself into a closet containing buckets, mops, and piles of brown soap. It was as easy as slipping into his own room off the Mendelssohnstrasse. Inside the closet, the fetid odors disappeared; here it was all disinfectant. He took a deep breath, held his nostrils until he grew used to the astringent smell. He found that he could just fit on top of the pile of buckets, his back resting against the severe metal wall, feet up, knees clasped passionately. That way he became half his real size. It was like sitting in the back of Lester and Clara's car, going to the opera. Listening for sounds on deck with exaggerated attention, he pulled out the blue vial and sniffed at it deeply. That soon wiped out the smell of the disinfectant. Then he tasted two drops, sucking at his finger. It was lovely. All around him he suddenly smelled powdered sugar, raisins, white wine, and opium. He sucked on his finger again and again. He dozed for a moment or two. Then, as he chewed on a cookie, four trumpets blasted above his head, sounding like a dozen. They were immediately followed by thumping kettledrums beating out the *Emden*'s repertoire of *fanfaren* marches. Shivers ran through the soles of Manfred's feet. The hair on his neck prickled. He clasped his knees tighter. He thought he could feel the wall behind him vibrate, he thought he could feel the whole ship vibrate. When it was over, a deathlike silence set in. No sound reached him. All sense of life on deck temporarily disappeared. Inside the closet, it was like sitting in a great metal vacuum, a black landscape without people or objects, waiting to be sucked out. Manfred swallowed hard. He had to remind himself where he was; he had to tell himself why. Then a few minutes later, without warning, everybody on deck began to sing "The Star-Spangled Banner." After they finished "The Star-Spangled Banner," they sang "Deutschland Über Alles." Manfred mouthed the words with them, sniffing at the blue vial, tasting another drop or two. The truth was that he was happy to hear the sound of human voices around him. Finally, all those on deck joined together on "Horst Wessel," singing very loudly. That must be the crew, Manfred thought, all those reedy tenors, square-shouldered baritones, giant basses. *Die fahne hoch, die Reihen dicht geschlossen* . . . After singing the whole "Horst Wessel," everybody went back to the beginning and repeated the first stanza, and then there was a huge cheer and the hollow sound of a cannon blasting away on deck.

By then Manfred could hardly keep his eyes open. While the guns were still echoing, his head fell forward and his nose began to run again. Inside the dank black box, a vision ecstatically appeared. He could clearly make out

Adele Gordon wearing her new formal, bought especially for B. P. Potter for tonight's party, showing it off for the family as they sat together in the living room after dinner. It was a white cotton dress with tiny cherry blossoms on it that left her shoulders bare. Two thin straps held the dress up. When she tried it on for them at home, all white and red and open flesh, Max had almost wept at the sight; Florence, too. How does it look, Adele kept asking, in her newly solicitous manner. Is it all right? Turning flirtatiously in front of them all, in the middle of the living room. You like it? she persisted. As though she didn't know, as though she didn't know everything. Manfred reached out for her in the dark then, wanting to tell her how wonderful she looked, how wonderful and appropriate, how right; but it was too late, he had waited too long, Adele had vanished and all the Gordons had vanished with her, Florence and Max, too, passed into limbo together before he could tell them he loved them. That was what he wanted to say and had forgotten. What might have been, the bleakest nostalgia of all.

He snivelled from dope into Benno's blue handkerchief, sitting still as a death's-head, immobilized in the dark on his strange metal perch. On deck now, the guests were beginning to depart. He could hear them trampling towards the gangplank; and he could tell that the crew were picking up their regular duties. Manfred grasped his knees tighter and held his breath. He must not give himself away. They were talking to each other right outside the door of his closet, just beyond the cave he had staked out for himself. They were arguing in heated German not ten inches from where he sat, clasping his knees ever-tighter to forestall panic. A swoop of blue and gold blinded Manfred. Familiar round flat scabbed faces were shouting at each other. Manfred's sphincter tightened unbearably; at the same moment he stained himself with a single drop of urine. Out in the ship's corridor, they were shouting dirty words in hard angry voices. Manfred put his hands over his ears. He did not want to hear what they were saying. He did not want to listen to obscenities. Obscenities made him think of his own death, of the grave.

An hour passed, more. Soon a profound shudder ran the length of the ship. Thre was a sudden rise in pitch from the ship's engines, just like the voices on deck a few hours ago; he was jolted against the wall, hard, then all was smooth again. The *Emden* began to move out of its berth in a silken line, executed with near-perfection, only the English could do it better, backing into Baltimore harbor, shredding the stinking refuse in its path with its double-screwed propeller. It wasn't until the ship reached Annapolis, an hour or so downstream, was in fact a couple of knots beyond the town, that Manfred was discovered asleep on his iron roost, snoring with tormented, strangulating rasps.

Chapter Nine

The officer in charge of the launch assigned to deliver Manfred Vogel to the Annapolis police was not especially grateful for having to slow the *Emden*'s progress downstream towards the Atlantic and home. In fact, he turned visibly irascible as the launch pulled away from the cruiser, which was bobbing with surprising force in the Chesapeake's nighttime swell, and kept showing his teeth to Manfred, who sat wrapped in a blanket in the stern of the small boat, shivering. The officer took his cues in everything from Captain Bachmann, naturally, and Captain Bachmann was still wearing his heavy metaphysical face tonight. Captain Bachmann was worried about the publicity that might come with the evening's events, specifically about possible charges of kidnapping. Bruno Hauptmann was on his mind; the matter had endless implications, if you began to pursue it, some of them too awful to contemplate. The officer-in-charge had not yet thought them through; they had not even occurred to him; he considered that he knew a stowaway when he saw one, and he assumed that the rest of the world would, too. The remainder of the crew had had its own morale punctured in the preceding hour by their captain, who was not one to hide his rage when aroused. Quite the contrary. An electrified alarm had coursed through the ship at the news of Manfred's discovery. The *Emden* had ground to a halt in midstream, its engines swelling in volume soon after Manfred was lifted out of his closet and brought before the ship's officers. An intense question-and-answer period immediately followed, led by Captain Bachmann in ferocious German. Manfred was not forthcoming. Somewhat stunned, he sat through the inquisition in silence, looking miserably at the floor. Everything had failed. There would be no escape from America, no reunion in Frankfurt, he would have to go on waiting in exile. As Captain Bachmann droned on, Manfred felt as though he was shrinking to the size of a dust mote. He kept clenching his teeth to make sure he would not give himself away. The grilling continued in loud, angry voices, unsuccessfully. Then, at the Captain's shouted orders, the launch was set overboard. Brusquely, the officer-in-charge and his crew of four helped Manfred down the side of the *Emden*. They made sure that he stepped lively, gripping him tightly by the upper arm. Their grip hurt. The April wind whipped neatly in from the western shore. A little ways to starboard the lights of Annapolis shone clearly. It was cold on the water. "Hurry," the officer ordered, angrily biting his lips in the swaying boat. He had been sipping a superb twenty-year-old Asbach brandy, glad to be on his way to Wilhelmshaven or wherever the *Emden* was headed, when Manfred's closet door had swung open to reveal him to an astounded seaman first-class who happened to be passing by. It seemed a shame and worse, a national humiliation perhaps, to conclude such a cordial (and useful) visit to a city with seventy-six German societies in its midst with such an unhappy incident.

After receiving the very stimulating radio message from the German ship, the Maryland cops took delivery of their hostage at the Annapolis waterfront. Certainly, nothing like this had ever happened to them before. The officer in charge of the launch made them sign a receipt for Manfred and, still in a rage, was off as soon as his boat could be backed away from the dock. The whole exchange took less than five minutes. The young man seemed woozy and somewhat hungover when the cops first saw him—drunk, was the initial diagnosis—but as the hours passed in the station house, while they all waited for Max Gordon to show up, the young man seemed to be making a swift recovery. "I'm OK," he kept saying, trying not to burst into tears. "No, thank you,"—to an offer of coffee—"I'm all right." He was still shivering, and he had to keep blowing his nose.

"Just what were you trying to prove out there anyway?" the sergeant at the desk wanted to know.

"I got lost," Manfred said. The answer seemed to cover all possible questions.

"Wait'll your folks get their hands on you," the sergeant muttered, turning away to answer a phone-call.

An hour later, Max walked in. "Oh, my God," he said, catching sight of Manfred sitting against the wall in his Indian sweater, knickers, and windbreaker. Manfred turned his head away. "What's going on anyway?" Max asked, but he spoke softly.

"I got lost," Manfred said.

"Lost," Max said. He stood over Manfred, smoking again. He kept puffing at his cigarette at every other word. "We've been going crazy on Fairfax Road worrying about you. Florence is out of her mind, in her condition." He paused a moment to blow smoke overhead. The sergeant sat at his desk, his thumbs tucked under his suspenders, listening to their exchange. "You got lost," Max said, shaking his head.

Manfred stared at his feet.

"Are you all right?" Max asked, sighing. "Did you fall in or anything?"

The sergeant laughed.

"I'm fine."

"Then let's get out of here. I don't like police stations."

Driving up the Annapolis Boulevard, Max was mostly quiet. "Your eyes looked awfully bloodshot back there," he finally said.

"It was from lack of air."

They drove on a few blocks. "You want to level?" Max asked.

"Level what?"

"The cop said you had twenty bucks on you. He said you had extra clothes, too. What does that mean?"

"Nothing. I wanted to buy a couple of presents."

"He said you had some medicine, too."

"That's for my stomach. I brought it from Germany."

"How come the whole world gets off the ship in time and you get lost in a broom closet?"

"It wasn't a broom closet, it was an accident. I shouted and nobody heard me."

"Manfred."

There was a silence. Cigarette ash dropped on the floor of the Buick. "Manfred," Max said again.

They were already halfway home. The moon was up over lower Baltimore. To the east the Union Trust tower shimmered in a silver light. Straight ahead the Bromo-Seltzer campanile, a beloved fraud, showed its huge clockface and the Bromo-Seltzer bottle on top. A few alert pedestrians walked the streets; it was Saturday night. Off on Manfred's right, a block away, were Hutzler's, Hochschild's and Stewart's. Across the street was the May Company, safely in the hands of Potter père. Gutman's was nearby, Brager-Eisenberg, Schleisner's, a feast of department stores, symbols of straining prosperity. The 31 streetcar ran on Eutaw Street, on the tracks in front of the Buick, the 32 on Howard. The Little Theater showed foreign films on Howard Street. There was a food market on Lexington Street, just behind them, acres of pungent shellfish, fresh vegetables, and other good things under one roof, and a half-mile away the Peabody Conservatory of Music lay at the feet of the ever-ready George Washington. Forest Park lay elsewhere. "You know," Max said, as they headed north and west, "it's going to be all right."

Manfred shrugged the remark off and moved closer to the door in the front seat.

"You don't believe it now, but it will," Max said. "I mean this alliance, this partnership, well, maybe it can't be called a family, you know what I mean, but it can be good enough, it can be a lot. I mean, given what there is. You know? It'll be all right. You'll see. I know what I'm talking about. I'm not as dumb as you think."

Fumble-mumble, Manfred thought, with the quickness of bad habit, but as the Buick stalled, then shuddered on, he began to feel something entirely different. More ash fell on the floor. "Your cigarette," Manfred scolded, yanking at the ashtray. If there was a mess in the car, Florence would start yelling at everybody. Max inhaled deeply and stubbed out his cigarette. A couple of broken smoke rings clouded the windshield. Manfred could see tiny bristles emerging on Max's chin and along the jawline. It was his evening's growth. His white hair was unkempt, as though he hadn't had time to comb it before heading for Annapolis, the index finger on his right hand was nicotine-stained.

Manfred turned to roll down the window to get rid of Max's smoke. The chilly air poured into the car. "Aren't you freezing?" Max asked. Manfred didn't answer. He was already beginning to plan a long weekend with the Ottingers, out on Rogers Avenue. The prospect had a certain lustre. The Ottingers were his rich friends. Out there, away from Fairfax Road for a while, he would have Lester's victrola to himself, a little *Tristan*, perhaps, a little *Rosenkavalier*, some heady French and Italian stuff, all in the original. Rhea would cook breakfast with a shaking hand, paralysis agitans, Shorty would serve it. From Stewie's bedroom there would be a smashing view of thoroughbred horses racing each other at dawn, beautiful dark shadows heading around the turf in the new sun; and adjoining Stewie's room, positioned in stunning silence, a bathroom as big and palatial as that of the Prince of Wales.

The Buick roared on. They had already reached Liberty Heights. Manfred glanced at Max again. His stubble seemed to have grown about half an inch. His right hand still rested on the wheel. He drove in silence, except for a vague purr that escaped him every now and then. Max was satisfied. He had done his job. He had offered Manfred a parent's consolation, spoken in a pure American voice. Even with its hesitations, its reticence and characteristic shyness, it had offered a certain amplitude. Consolation and hope were at the center of Max's words, they were its very heart. And there was also that thing (which Manfred clearly heard) that said, without guile: this way for now and no other. It was hardly a thunderclap, but it carried exactly the right tone. It projected dead center. In any case, Manfred Vogel was sick to death of the agitated sound of other people's overbearing, overheated voices.

"For God's sake, close the window," Max said. His teeth were chattering. Manfred did as he was told. He felt no sense of resistance. He welcomed Max's order. Things were probably as they should be. He couldn't wait to get home and have the fuss over with. There would be plenty of emotion. Florence would be all over him, in her sizzling uncontrollable way, Adele worshipping and remorseful at his feet (that is, if she was feeling like it, if the mood struck her). That would be his reward for his adventure, adoration and love, and a certain pleasant notoriety, as well.

"German Boy Grins Real Grin At Last," he read to himself in the streetlights that lined the side of the road, as they drove on. Max had begun to race the trolley down Liberty Heights. "Frantic Driver Jailed for Speeding..."

Part Four

A few years later, in June, 1939, a letter Manfred had mailed to the Mendelssohnstrasse was returned to him with the words Address Unknown stamped on the envelope in red ink. *Adresse Unbekannt.* At the sight of the German words, angled at forty-five degrees, Manfred let out a little animal-like scream without knowing it. A headache began, near-panic took over. Standing there in the Gordon hallway, he was unable to think, unable even to remember his own name for the moment. For once, witnessing this brief scene, the household was silent. No one spoke. They all left Manfred alone. Even Dickie, who was the baby, was quiet. Dickie was very susceptible to the moods of his parents and to Manfred's as well. Manfred had just arrived home from Columbia University, where he was a freshman (to Athol Burns's reasonable, if not total, satisfaction). Dickie's sister, at the moment, was on her way south from Bennington for summer vacation, after finishing her junior year. Fairfax Road was behind them. Rowhouses, tin garages, and touch football in the street no longer existed. Neither did Billy Brent, Laura Piscitelli, or anyone like them. It was the way Max had always planned it. They lived now in nine rooms in a fieldstone house on Bancroft Road off Park Heights, out in the country, among the other well-to-do Jews.

When his letter arrived back in Baltimore, the stamped words somewhat blurred from handling, Manfred instantly knew that there was no mistake. His parents were really gone from Frankfurt, or worse. By worse, he was not quite sure what he meant. There were so many possibilities that did not bear thinking about. The newspapers were full of them. That was why Manfred had screamed without thought, making a sound that had caught like a fishhook in his throat (and that Max, Florence, and Dickie Gordon would never forget). After that, after months of futile attempts to trace his parents, with Kurt's passionate involvement in Cambridge and Uncle Oskar's in Zurich, nothing was ever discovered about Elsa and Julius Vogel.

They had disappeared into a void, without clues. In September, war in Europe began. The line to Germany was broken. Another world began to take shape. Manfred was then nineteen.

The terrible facts had to be faced and faced again and faced still another time. Elsa and Julius Vogel had disappeared like all the others (sooner or later): Selig Vogel and his child-bride Ernestine into Bergen-Belsen; Jenny the voluptuous war-widow, hidden for three silent years in the storage room of her faithful dressmaker's Frankfurt apartment, along with a half-dozen ancient clothing-dummies with figures exactly like Jenny's; and Benno Mann, the family pederast, become a slave in a Silesian munitions factory, marked by a pink Star of David sewn on the sleeve of his jacket by his own hand. Of these six, Vogels and Manns, only Jenny was to die a natural death, or a death that was confirmable as natural, seized by heart failure in Haifa in 1956 while visiting her daughter Lisl in her new apartment overlooking the Mediterranean.

As suggested above, Oskar and Lotte Mann, the family worldlings, escaped from Germany after buying their way out of the homeland in 1940, finally settling in Zurich, where they have since lived, comfortably enough, on the interest of their Swiss bank accounts (the very bank accounts Manfred and Kurt are to inherit). But the Manns are no longer rich, postwar inflation having done its job as effectively in Zurich as everywhere else. Nor do they travel as they used to. These days they mostly sit at home in their four-room condominium and watch television together. They are in bed by ten-thirty. They have become Swiss hearth-lovers. Neither expects to live much longer, in any case, although they do not talk about it. Lotte has severe hypertension, a chronic lifetime ailment. Her voice is as throaty as ever, her bony face as intelligent; she now wears a hearing aid. Lotte writes to Manfred and Kurt several times a year without fail, humorous, chatty, mimeographed letters for the two of them, full of real news, real gossip, insofar as Zurich supplies real gossip, and real intermittent talk about the Arts (Oskar and Lotte still depend on Art for consolation). A reception for Leonard Bernstein, a concert by Boulez, a visit to Zurich by the latest Russian émigré pianist, such matters. Oskar suffers from senile dementia, which comes and goes. He has lost all his teeth since V-E Day.

In America, of course, the Gordons have become rich. Marriages are made in heaven, but engagements are made at M. Gordon & Son and at a heady rate, as Max always believed they would be. Three stores in Baltimore, one in Washington, another in Richmond, Virginia, a golden chain, so to speak. In the process of becoming rich, Florence cooled off to a considerable degree and discovered a kind of reserve in herself, a reticence, that she had not known existed. The dependable presence of money helped to do that; it gave her confidence and calmed her energy. Her friends now sit in her living room on Bancroft Road, just as she once sat in Clara's, and nod approvingly at what she has to say. Max, on the other hand, has become somewhat stuffy, a surprise to some people, developing the innocent habit with time of pontificating laboriously on almost any subject that comes up. It is because he really believes in his own success, in the justice of it. Max is full of unquestioned authority on every matter and a recognized leader of the powerful Jewish community in Baltimore. Over the years, he has also put on thirty pounds and more, which has not, however, entirely affected his facetious habits. Bad puns are still his custom and his pleasure. Max's hair, if possible, is whiter, more beautiful than ever. Both Max and Florence have enjoyed Bancroft Road. Their house, which is one of a kind, sits on a small hill from which they can overlook the rest of the world. It seems almost enough.

Their son Dickie was as precocious as Manfred and Adele. He spoke in full sentences at the age of twelve months and played the piano at three, astonishing everyone. A Mozart, Clara suggested, half joking, and for a moment, Max thought it a possibility. (It wasn't.) Dickie has always thought of Manfred as a real brother and still worships him as something of a hero. Naturally, this pleases Manfred and Max, too. (There are sixteen years between Manfred and Dickie.) Everyone like Dickie. He is friendly, straightforward, and sometimes witty, a great asset to M. Gordon & Son, which he now runs successfully from the greatly enlarged Lexington Street store. Dickie Gordon has virtually no memories of Fairfax Road or of being poor. In a real sense, he grew up in a different family, another Gordon clan, one with new hopes and different dreams, his alone. Manfred and Adele have to remind themselves of that from time to time.

Adele herself, after graduating from Bennington, decided to become a modern dancer, against her parents' wishes. This experiment lasted for three months, during which she and her mother did not speak to each other (not for the first time). Then Adele taught nursery school at Montessori in New York for a semester. Another experiment. She has also written some poetry.

One poem was published in the *Prairie Schooner*, causing great excitement among the Gordons and their relatives, especially Clara and Lester. (Florence had to ask what a prairie schooner was.) Today she has a master's in Social Work and specializes in the problems of black families. Adele remains Adele, as no doubt she always will. Life pushes hard, but she has always known how to push back, without complaints. She is almost exactly the same, with only slight modulations in tone, the old homing instinct once again in place, for good. Adele married an internist from Shaker Heights, Cleveland, where she now lives, mother of four very bright children. (The youngest suffers from muscular dystrophy.) Together, they all sometimes share bird-watching expeditions, to Mexico and other places. Adele's husband is very bright, too, and generally speaking a patient and compassionate man. He is also successful, like all doctors. He understands how to deal with his wife, how to make her happy; and he does (although there are moments). Adele and Manfred speak to each other every month by phone.

As for the Ottingers, they go on in much the same way out in the big house in Mt. Washington. Both Clara and Lester are slightly shrunken by time. Clara's massiveness is somehow diminished, and Lester seems more and more bored with the life around him. The Metropolitan Opera no longer visits Baltimore, the Hudson River paintings now hang in the Baltimore Museum of Art, and many old friends, formerly snubbed, have died. These days everything seems to be hair-in-the-soup. They have been unable to give up the big house or the view of Pimlico across the road, even though the neighborhood has deteriorated seriously over the years. Lester has grown a bit listless is the consensus (Babs and Stewie both agree), and his drinking is an unfortunate ongoing issue in the family. (Babs and Stewie are contentious, even rancorous, on the subject.) But although shrunken, Clara is as purposeful, as steady as ever. She still makes all the rules and sees to it that they are enforced.

Eventually, over the course of two decades and more, Manfred himself went on to become a worldly success, after service in U.S. Army Intelligence from 1942 to the end of 1945. (So did Kurt Vogel in Cambridge, on a different scale: academic, scientific, local.) Manfred's career was built around aspects of language and diplomacy. It turned out that he was expert at both. As the career expanded, he achieved a reputation for brilliance and eloquence among his colleagues and a considerable degree of national recognition. At one point, he was a senior adviser to a Democratic administration, charged with delicate foreign policy matters, consulted for the judiciousness and scope of his opinions. His essays have appeared in distinguished journals in both the U.S. and England. They are now being anthologized for book publication. (No one on Fairfax Road or Rogers Avenue or at Forest Park High was very surprised at any of this.) Obviously, it has been infinitely better than Frankfurt, this strange foreign life he stumbled into, more or less, with its strange foreign aura with which he is still not wholly at home, more generous, more giving, more open than anything he might have known in the old Europe, even in the best of times. He has rarely—with a few exceptions, of course—had many doubts about that. In the end, Manfred keeps telling himself, thinking of his wife, once one of the adorable Baltimore dollies, but an exceptional idiosyncratic one, full of fresh intelligence on which Manfred unresistingly depends; of their only child, his own son, now grown to reasonably splendid manhood (one of the old dreams, one of the best); thinking too of his restless brood of jealous ghosts, all those possessive Vogels and Manns greedily clinging to him like a second skin, bitter source of never-ending sadness and incurable silent depressions in both Manfred and his brother—in the end, he thinks, the world sometimes makes the attempt to even things up. As an American, habitually folding his grief inside himself like a furled flag, he tells himself this without irony. And so he tells his brother, without irony, as they correspond with each other across the Atlantic. As an American, Manfred really believes it. In America, balances are struck, proportions eventually righted. What else, he asks Kurt, makes America different? And in fact the conviction has worked for him over the years, in the face of everything, considering everything that might have been. That is, it has held (to hold: *behaupten, aushalten, ausdauern, reichen*, et al—see Cassell's German-English Dictionary, revised and considerably enlarged by Karl Breul, Cambridge University Reader in Germanic, 1906). It has been like a gift, he thinks. It has been like a gift of time to a favored child.

Design by David Bullen
Typeset in Mergenthaler Granjon
with Ehrhardt display
by Wilsted & Taylor
Printed by Maple-Vail
on acid-free paper